INNOVATIVE DEVELOPMENT

Global Hawk AND DarkStar

Flight Test in the HAE UAV ACTD Program

2

Jeffrey A. Drezner

Robert S. Leonard

Prepared for the United States Air Force

RAND

Project AIR FORCE

The research reported here was sponsored by the United States Air Force under Contract F49642-01-C-0003. Further information may be obtained from the Strategic Planning Division, Directorate of Plans, Hq USAF.

ISBN: 0-8330-3113-9

Cover photo courtesy of Northrop Grumman Corporation.
www.northropgrumman.com
Reprinted by permission.

Cover design by Barbara Angell Caslon

Published 2002 by RAND
1700 Main Street, P.O. Box 2138, Santa Monica, CA 90407-2138
1200 South Hayes Street, Arlington, VA 22202-5050
201 North Craig Street, Suite 102, Pittsburgh, PA 15213-1516
RAND URL: http://www.rand.org/
To order RAND documents or to obtain additional information, contact Distribution Services: Telephone: (310) 451-7002;
Fax: (310) 451-6915; Email: order@rand.org

PREFACE

The High-Altitude Endurance Unmanned Aerial Vehicle (HAE UAV) Advanced Concept Technology Demonstration (ACTD) program incorporated a number of innovative elements into its development strategy. As a condition of conducting this ACTD, Congress required that an independent third party study its implementation. RAND was chosen for this role and has been following the HAE UAV ACTD program since its inception.[1]

The flight test program and user demonstration are core components of the HAE UAV ACTD program. These are the activities through which the ACTD program objective—demonstrating military utility through early user participation and test—is accomplished.

The initial phases of our research were sponsored by the Defense Advanced Research Projects Agency (DARPA); the current research was sponsored by the U.S. Air Force. The core objective of the research was twofold: to understand how the innovative acquisition strategy used in the HAE UAV ACTD program has affected program execution and outcomes, and to draw lessons from this experience that would be applicable to the wider acquisition community.

---

[1]See Geoffrey Sommer, Giles K. Smith, John L. Birkler, and James R. Chiesa, *The Global Hawk Unmanned Aerial Vehicle Acquisition Process: A Summary of Phase I Experience*, MR-809-DARPA, Santa Monica: RAND, 1997; and Jeffrey A. Drezner, Geoffrey Sommer, and Robert S. Leonard, *Innovative Management in the DARPA High Altitude Endurance Unmanned Aerial Vehicle Program: Phase II Experience*, MR-1054-DARPA, Santa Monica: RAND, 1999.

This report addresses flight test and user demonstration outcomes and issues relevant to the HAE UAV ACTD program. In it, we assess the extent to which the innovative acquisition strategy used in this effort affected the conduct of the flight test program. This report is one of three supporting documents resulting from the current research effort; the other two documents address the activity content of the program and issues associated with transition management. A separate executive summary draws broad lessons from the HAE UAV experience as a whole.

This research was sponsored by the Global Hawk System Program Office (GHSPO), part of the Aeronautical Systems Center's Reconnaissance Air Vehicle (ASC/RAV) office in the Air Force Materiel Command (AFMC). It was conducted within RAND's Project AIR FORCE.

Reports in this series are:

MR-1473-AF, *Innovative Development: Global Hawk and DarkStar—Their Advanced Concept Technology Demonstrator Program Experience, Executive Summary,* Jeffrey A. Drezner, Robert S. Leonard

MR-1474-AF, *Innovative Development: Global Hawk and DarkStar—HAE UAV ACTD Program Description and Comparative Analysis,* Robert S. Leonard, Jeffrey A. Drezner

MR-1475-AF, *Innovative Development: Global Hawk and DarkStar—Flight Test in the HAE UAV ACTD Program,* Jeffrey A. Drezner, Robert S. Leonard

MR-1476-AF, *Innovative Development: Global Hawk and DarkStar—Transitions Within and Out of the HAE UAV ACTD Program,* Jeffrey A. Drezner, Robert S. Leonard

PROJECT AIR FORCE

Project AIR FORCE, a division of RAND, is the Air Force federally funded research and development center (FFRDC) for studies and analyses. It provides the Air Force with independent analyses of policy alternatives affecting the development, employment, combat readiness, and support of current and future aerospace forces.

Research is performed in four programs: Aerospace Force Development; Manpower, Personnel, and Training; Resource Management; and Strategy and Doctrine.

CONTENTS

FIGURES

TABLES

SUMMARY

The United States has seen a three-decade-long history of poor outcomes in unmanned aerial vehicle (UAV) development efforts. Technical problems have led to cost growth and schedule slip as well as to disappointing operational results. Costs have tended to escalate so much during development that the resulting systems have cost more than users have been willing to pay, precipitating program cancellation in almost every case. This history prompted the unique developmental approach adopted at the beginning of the High-Altitude Endurance Unmanned Aerial Vehicle (HAE UAV) program.

There has also been a long history of efforts made to improve the efficiency and effectiveness of weapon system acquisition policy, processes, and management. Capturing the experience from ongoing or recently completed efforts employing nonstandard or innovative acquisition strategies can facilitate such improvements. This research contributes to that effort.

In 1994, the Defense Advanced Research Projects Agency (DARPA), in conjunction with the Defense Airborne Reconnaissance Office (DARO), began the development of two UAVs. These systems were intended to provide intelligence, surveillance, and reconnaissance information to the warfighter. As such, they responded both to recommendations made by the Defense Science Board and to operational needs stated by DARO on behalf of military service users.

With congressional support, DARPA adopted an innovative acquisition strategy that differed from normal DoD procedures. The strategy's innovations are embodied in seven specific elements: designation as an Advanced Concept Technology Demonstration (ACTD) program; use of Section 845 Other Transaction Authority (OTA); use of Integrated Product and Process Development (IPPD) and a management structure based on Integrated Product Teams (IPTs); contractor design and management authority; a small joint program office; user participation through early operational demonstrations; and a single requirement—unit flyaway price (UFP)—with all other performance characteristics stated as goals.

The HAE UAV ACTD program included two air vehicles: a conventional configuration and a low-observable (LO) configuration. A common ground segment (CGS) was added not long after program initiation. The ACTD program was structured into three phases. Phase I was a design competition for the conventional Tier II+ system. Phase II included the development and test of both the Tier II+ (Global Hawk) and the LO Tier III– (DarkStar). Phase III involved the demonstration and evaluation (D&E) activity leading to a military utility assessment (MUA).

RAND has been assessing the execution of the HAE UAV ACTD program's innovative acquisition strategy since the program's inception in 1994. Previous reports have documented the effects of that strategy on Phase I and Phase II of the ACTD program.[2] The current research addresses the completion of Phase II, the transition to Phase III, and the transition to post-ACTD activities. This report specifically addresses the ACTD flight test program. Two companion documents (MR-1474-AF and MR-1476-AF) describe transition issues and outline program activity content. A separate executive summary (MR-1473-AF) presents our overall assessment of the acquisition strategy and suggests improvements.

Excluding the two initial DarkStar flights in early 1996, the HAE UAV ACTD flight test program was fairly cautious. A pattern of learning—test, analyze, fix, and test—is clearly evident for both Global Hawk and DarkStar. Master test plans were developed that included detailed objectives for each engineering test flight. The MUA process

---

[2]See Geoffrey Sommer, Giles K. Smith, John L. Birkler, and James R. Chiesa, *The Global Hawk Unmanned Aerial Vehicle Acquisition Process: A Summary of Phase I Experience*, MR-809-DARPA, Santa Monica: RAND, 1997; and Jeffrey A. Drezner, Geoffrey Sommer, and Robert S. Leonard, *Innovative Management in the DARPA High Altitude Endurance Unmanned Aerial Vehicle Program: Phase II Experience*, MR-1054-DARPA, Santa Monica: RAND, 1999.

was reasonably well documented both in planning and in execution, but some detail, including exit criteria and priority setting, was missing from the plans. Moreover, limited personnel and funding resources constrained flight test execution.

DARKSTAR

The first DarkStar flew in March 1996 and crashed on takeoff on its second flight attempt in April 1996. The second DarkStar air vehicle first flew 26 months later and flew only five times, accumulating six flight hours before the program was terminated in January 1999. DarkStar did not participate in D&E flight test activity. Thus, very little can be concluded from its limited flight testing. Our reading of DarkStar test results suggests that the performance of the air vehicle would have been considerably less than the stated goals, but any prediction of ultimate mission performance would be highly uncertain.

Not enough flight experience was accumulated to allow for an understanding of the flight characteristics of DarkStar. Unexplained oscillation problems during flight were not resolved. The performance of the ground segment was not determined. DarkStar sensor payloads apparently performed well in tests but were never flown onboard the DarkStar air vehicle.

GLOBAL HAWK

Table S.1 summarizes the Global Hawk flight test program by phase and air vehicle. The first flight was in February 1998. The first air vehicle was clearly the workhorse of the program, participating in both Phase II and Phase III. The third through fifth air vehicles participated only in Phase III. Although shorter in calendar length and flying fewer hours than initially planned, the flight test program did accumulate enough experience to demonstrate Global Hawk's military utility.

Six outcomes of Global Hawk's ACTD flight test experience are either partially or wholly attributable to the program's novel acquisition approach:

Table S.1

Summary of the Global Hawk Flight Test Program by Phase and Air Vehicle (number of sorties/flight hours)

Phase	Air Vehicle 1	Air Vehicle 2	Air Vehicle 3	Air Vehicle 4	Air Vehicle 5	Total
II	12/102.9	9/55.1				21/158.0
III	13/225.4		9/121.8	11/167.8	4/39.0	37/554.0
III+					5/25.1	5/25.1
Total	25/328.3	9/55.1	9/121.8	11/167.8	9/64.1	63/737.1

- The mission planning process was cumbersome and time-consuming, affecting the pace of the flight test program as well as the workload of flight test personnel. The contractors knew at the time of the Phase II bid that significantly more funds would be required to develop a mission planning system suitable for sustained operations. However, because the focus of the ACTD was on demonstrating military utility, which at the time was not well defined and did not specify timely sortie generation, a conscious decision was made not to make this investment. Had mission planning been made a priority (i.e., incorporated into a definition of utility early in the program), more funding might have been committed to it, albeit at the expense of other activities.

- The program lacked sufficient resources in terms of both personnel and spares. This paucity of trained personnel and spares limited the number of vehicles that could be flown at any one time during the latter portion of the flight test program. This was attributable in part to the reallocation of resources within the program to cover increased nonrecurring engineering activity, and in part to a highly constrained budget throughout the duration of the ACTD.

- The pace of the flight test program was fast given its cumbersome mission planning process and limited resources. Test personnel were clearly overburdened, which appears to have been a contributing factor in the taxi mishap of air vehicle 3.

- The contractors were designated the lead for flight test program execution, with the government program office assuming responsibility for liability and contingency planning. Contractors thus held significantly more responsibility than is the case in traditional programs, creating a somewhat different program management dynamic. Yet contractors do not have the necessary capabilities, experience, and perspective (culture) to run all aspects of a test program. The government test and operational communities thus took on a large portion of the planning and execution of the flight test program. Their assistance was essential to the accomplishments of the program.

- ACTDs are specifically intended to explore innovative concepts of operations (CONOPS) throughout the D&E phase of the program. Yet differences in perspective between the ACTD and post-ACTD user communities regarding CONOPS proved to be a serious impediment to the program's transition into the Major Defense Acquisition Program (MDAP) process. The initial CONOPS was generated by the DARPA joint program office and was then modified and expanded by the Joint Forces Command (JFCOM) as part of its responsibility as the designated ACTD user. The post-ACTD CONOPS of the Air Combat Command (ACC) is similar to current systems in terms of its access to sensor retasking and dissemination pathways. JFCOM's CONOPS takes advantage of advances in communications and processing technology and adopts a joint orientation. The ACTD demonstrated the JFCOM CONOPS; ACC has not demonstrated its CONOPS with respect to Global Hawk.
- ACTDs do not have approved operational requirements; they are intended to help refine operational requirements through lessons from flight test experience. Differences in operational requirements definition between ACTD and post-ACTD users also inhibited the program's transition to the MDAP. The extent to which the capabilities of the ACTD configuration should determine the requirements for a post-ACTD system is the underlying issue. The spiral development concept planned for use in post-ACTD development implies that requirements will evolve as the system's configuration evolves via block upgrades. As a result of this process, early configurations will not have the full capability that ACC, the force provider, desires. Nevertheless, the ACTD D&E phase did provide information critical to developing an operational requirement for the MDAP program.

The performance of Global Hawk was close to predicted goals but fell short in several significant areas. Empty weight increased 16 percent, and lower-than-predicted aerodynamic performance resulted in a 24 percent endurance shortfall (32 hours versus 40 hours) and a 7.7 percent shortfall in mission cruise altitude (60 kft versus 65 kft).[3] The program demonstrated the feasibility of autonomous flight and long endurance at altitude, and most of the communications and data links were demonstrated sufficiently. The synthetic aperture radar (SAR) sensor appears to provide high-quality imagery. However, the program did not demonstrate CGS control of multiple vehicles; nor was the electro-optical/infrared (EO/IR) sensor characterized sufficiently.

Some participants believe that neither the content of the flight test program (what was done) nor the approach used (how it was done) was greatly affected by the acquisition strategy. Evidence suggests that the dominant influence on the test program was the nature of the system. Until the HAE UAV ACTD program, very little experience had been accumulated with large autonomous UAVs. System characteristics determined the pace of the program, the profile in which flight hours were accumulated over time, and the scope of envelope expansion testing. The acquisition approach did, however, influence some key elements of the test program: the increased contractor responsibility for test program planning, direction, and execution; the early operational testing in the form of user demonstrations; and the explicit exploration of operational concepts and requirements.

One important lesson from the flight test program was the necessity for early involvement on the part of operational users—in this case the 31st Test and Evaluation Squadron (TES). The operational users not only provided much-needed skills and capabilities in support of the test program but also introduced a critical operational perspective into the conduct of the flight test program.

---

[3]The original DARPA mission profile shows a 3000-nm ingress, a 24-hour on-station segment at 65 kft, and a 3000-nm egress. It is this on-station "cruise" segment that Global Hawk cannot achieve. Global Hawk can achieve an altitude of 65,000 ft for shorter periods of time under certain environmental and weight-related (e.g., fuel remaining) conditions.

ACKNOWLEDGMENTS

This research would not have been possible without the cooperation of officials associated with the HAE UAV ACTD program in the U.S. Air Force, the Office of the Secretary of Defense, DARPA, and industry. Special thanks are due to the government program office personnel and to the contractors who provided information and spent considerable time with us discussing the HAE UAV ACTD program.

We would also like to extend our thanks to Geoffrey Sommer, ex officio project team member, who provided suggestions and observations during the research as well as an unofficial review of the draft reports. Our formal technical reviewers, Giles Smith and Frank Fernandez, provided excellent reviews of the draft reports.

In addition, we would like to thank Natalie Crawford and Timothy Bonds in Project AIR FORCE for providing their time and resources to ensure that this research was completed to the highest-quality standards.

Any remaining errors are the sole responsibility of the authors.

ACRONYMS

ACC	Air Combat Command
ACC/DO	Air Combat Command/Director of Operations
ACTD	Advanced Concept Technology Demonstration
AFFTC	Air Force Flight Test Center
AFMC	Air Force Materiel Command
AFMSS	Air Force Mission Support System
AFOTEC	Air Force Operational Test and Evaluation Center
ASC/RAV	Aeronautical Systems Center/Reconnaissance Air Vehicle
ATC	Air traffic control
ATD	Advanced technology demonstrator
AX	Attack Aircraft Prototype [program]
CAS	Close air support
CAX	Combined Arms Exercise
C4I	Command, control, communications, computers, and intelligence
C4ISP	Command, control, communications, computers, and intelligence support plan
CCO	Command-and-control officer
CDL	Command data link
CGS	Common ground segment
CINC	Commander in chief
CONOPS	Concept of operations; concepts of operations

CONUS	Continental United States
D&E	Demonstration and evaluation
DARO	Defense Airborne Reconnaissance Office
DARPA	Defense Advanced Research Projects Agency
DESA	Defense Evaluation Support Activity
DT/OT	Development test/operational test
DT&E	Development test and evaluation
EAFB	Edwards Air Force Base
ECS	Environmental control system
EMD	Engineering and manufacturing development
EO/IR	Electro-optical/infrared
FAA	Federal Aviation Administration
FL	Flight level
FOT&E	Follow-on operational test and evaluation
FSD	Full-scale development
GHSPO	Global Hawk System Program Office
GMTI	Ground moving target indicator
HAE UAV	High-Altitude Endurance Unmanned Aerial Vehicle [program]
IAP	Integrated assessment plan
IG	Inspector General
IMMC	Integrated mission management computer
INS	Inertial navigation system
IOT&E	Initial operational test and evaluation
IPPD	Integrated Product and Process Development
IPT	Integrated Product Team
ISR	Intelligence, surveillance, and reconnaissance
ISS	Integrated sensor suite
JEFX	Joint Expeditionary Forces Experiment
JFCOM	Joint Forces Command
JMETL	Joint mission-essential task list
JPO	Joint program office

JTF	Joint Task Force
JTFEX	Joint Task Force Exercise
LCRS	Launch, control, and recovery station
LMSW	Lockheed Martin Skunk Works
LO	Low observable
LRE	Launch and recovery element
LWF	Lightweight Fighter
MAE	Medium-Altitude Endurance
MAR	Monthly acquisition report
MCE	Mission control element
MDAP	Major Defense Acquisition Program
MOB	Main operating base
MUA	Military utility assessment
NAS	National air space
NTC	National Training Center
OA	Operational assessment
O&S	Operations and support
OM	Operational manager
OPLAN	Operations plan
ORD	Operational requirements document
OT	Other Transaction
OT&E	Operational test and evaluation
OTA	Other Transaction Authority
RTB	Return to base
RTO	Responsible test organization
SAR	Synthetic aperture radar
SATCOM	Satellite communications
SIL	Simulation and Integration Laboratory
SMM	System maturity matrix
SPO	System Program Office
TAUV	Tactical Unmanned Aerial Vehicle [program]
TEMP	Test and evaluation management plan

TES	Test and Evaluation Squadron
TO	Takeoff
TPM	Technical performance measure
UAV	Unmanned aerial vehicle
UFP	Unit flyaway price
USACOM	United States Atlantic Command
WPAFB	Wright-Patterson Air Force Base

Chapter One

INTRODUCTION

In 1994, the Defense Advanced Research Projects Agency (DARPA), in conjunction with the Defense Airborne Reconnaissance Office (DARO), began the development of two unmanned aerial vehicles (UAVs). These systems were intended to provide intelligence, surveillance, and reconnaissance (ISR) information to the warfighter. They responded to the recommendations of the Defense Science Board and to operational needs stated by DARO on behalf of military service users.

UAV and tactical surveillance/reconnaissance programs have a history of failure in the United States resulting from inadequate integration of sensor, platform, and ground elements, together with unit costs far exceeding what operators have been willing to pay. This history of failure has contributed to a sense of frustration and to a realization that the DoD needs to explore ways to simplify and improve the acquisition process. To overcome these historical problems, DARPA, with congressional support, adopted an innovative acquisition strategy that differed from normal DoD acquisition procedures in several important ways. These innovations were embodied in seven specific elements of the strategy: designation as an Advanced Concept Technology Demonstration (ACTD) program; use of Section 845 Other Transaction Authority (OTA); use of Integrated Product and Process Development (IPPD) and a management structure based on Integrated Product Teams (IPTs); contractor design and management authority; a small joint program office; user participation through early operational demonstrations; and a single requirement—unit flyaway price (UFP)—with all other performance characteristics stated as goals.

The High-Altitude Endurance Unmanned Aerial Vehicle (HAE UAV) ACTD program consisted of two complementary system development efforts: the conventionally configured Tier II+ Global Hawk and the Tier III– DarkStar, which incorporated low-observable (LO) technology into the design of the air vehicle. The program also included a common ground segment (CGS) that was intended to provide launch, recovery, and control for both air vehicles.

The ACTD program was structured into three phases. Phase I was a design competition for the conventional Tier II+ system. Phase II included the development and test of both the Tier II+ (Global Hawk) and the LO Tier III– (DarkStar). Phase III involved the demonstration and evaluation (D&E) activity leading to a military utility assessment.

RAND has been analyzing the execution of the HAE UAV ACTD program's innovative acquisition strategy since the program's inception in 1994. Previous reports have documented the effects of that innovative acquisition strategy on Phase I and Phase II of the ACTD program.[1] The current research addresses the completion of Phase II, the transition to Phase III, and the transition to post-ACTD activities.

As the HAE UAV ACTD program transitioned to Air Force management and subsequently entered its D&E phase, it became useful to break down our analysis into several key issue areas: transition management issues; the activity content of the program; and the flight test program. This report specifically addresses the flight test program. A separately published summary document synthesizes the results of these three efforts; draws conclusions regarding the advantages and disadvantages of this innovative acquisition strategy; and suggests ways in which the strategy can be enhanced.

---

[1]See Geoffrey Sommer, Giles K. Smith, John L. Birkler, and James R. Chiesa, *The Global Hawk Unmanned Aerial Vehicle Acquisition Process: A Summary of Phase I Experience*, MR-809-DARPA, Santa Monica: RAND, 1997; and Jeffrey A. Drezner, Geoffrey Sommer, and Robert S. Leonard, *Innovative Management in the DARPA High Altitude Endurance Unmanned Aerial Vehicle Program: Phase II Experience*, MR-1054-DARPA, Santa Monica: RAND, 1999. See also Robert S. Leonard, Jeffrey A. Drezner, and Geoffrey Sommer, *The Arsenal Ship Acquisition Process Experience*, MR-1030-DARPA, Santa Monica: RAND, 1999.

OBJECTIVES

The process of improving acquisition management methods, policy, and supporting analyses requires the accumulation of experience from ongoing or recently completed projects, especially those involving unusual situations or innovative acquisition strategies. This research contributes to that effort. The objective of this research was twofold: to understand how the innovative acquisition strategy used in the HAE UAV ACTD program affected program execution and outcomes, and to identify lessons that might be applied to a wider variety of programs in order to improve DoD acquisition strategies.

This report addresses the HAE UAV ACTD flight test program. This flight test program, conducted for the most part from February 1998 to September 2000, includes both engineering development testing and user demonstrations. Engineering development testing was conducted predominantly during Phase II of the ACTD; Phase III of the ACTD consisted predominantly of the D&E activities. The goal of this report is to understand how the HAE UAV ACTD program's innovative acquisition strategy affected the execution of the flight test program.

RESEARCH APPROACH

This multiyear research effort tracked and documented the execution of the HAE UAV ACTD program through the completion of the ACTD. The overall project was organized into three tasks.

Task 1: HAE UAV ACTD Program Tracking

The primary research task was to track and document the experience of both the program office and contractors as the HAE UAV ACTD program proceeded. This task involved periodic discussions with both the government program office and contractors in efforts to understand current program status, key events and milestones, and how the innovative elements of the acquisition strategy were being implemented. Task 1 also involved a thorough review of program documentation, including solicitations, proposals, Agreements, memoranda, and program review briefings. Information on program funding, scheduling, and the flight test program was also reviewed.

Through discussions and reviews of documentation, we were able to assess whether the acquisition strategy was having the expected effect as well as to identify issues arising in the course of program execution that either affected or were affected by the acquisition strategy.

Task 2: Comparisons to Other Programs

In this portion of the research, we collected and analyzed historical cost, schedule, and flight test data from comparable past programs. Relatively little historical data has been preserved on past UAV programs at a detailed level, limiting their value as a baseline for comparison with the current HAE UAV ACTD programs. Therefore, we assembled data on program outcomes from broader databases of historical experience in order to assess HAE UAV ACTD program outcomes in a historical context. Additionally, we examined the transition experience and the adequacy of testing of other programs in order to gain insight on the relative utility of the strategy employed in the HAE UAV ACTD program.

Task 3: Analysis and Lessons Learned

In this task, we synthesized the information collected under Tasks 1 and 2 and developed two kinds of overall results. One focused on understanding the extent to which the HAE UAV ACTD program was implemented as planned and the degree to which the program had achieved its expected outcomes. The other focused on comparisons between the HAE UAV ACTD program and other comparable programs. Together, these tasks yielded an understanding of the strengths and weaknesses of the overall HAE UAV ACTD acquisition strategy. We then interpreted those results in terms of lessons that might be applied to future programs.

Chapter Two

OVERVIEW OF THE HAE UAV ACTD FLIGHT TEST PROGRAM

EVOLUTION OF THE INITIAL PLAN AND UNDERLYING PHILOSOPHY

The December 15, 1994, HAE UAV ACTD management plan, one of the earliest documents providing guidance for program execution, addressed testing in only the broadest terms. Flight tests were to be conducted as part of Phases II (development) and III (D&E). Phase II would include a 12-month test program using two engineering development UAV models and an engineering test model CGS.[1] Phase III, a 24-month field demonstration involving the participation of operational users, was to use field demonstration model UAVs: eight conventional HAE UAVs (Global Hawk), up to eight LO HAE UAVs (DarkStar) if funding permitted, and two additional CGSs. Phase III was to include the combined testing of the conventional and LO vehicles during field demonstrations with operational forces. The Phase II tests for both air vehicles were intended to be conducted by the contractors. Phase III testing included participation in DoD training exercises, as distinct from traditional operational testing.

The expectation in the December 1994 management plan was clearly that all air vehicles, both conventional and LO, would be essentially identical, leading to a direct and orderly transition into production. The plan stated that at the completion of the ACTD, a residual opera-

[1]See Defense Adanced Research Projects Agency, *HAE UAV ACTD Management Plan*, Arlington, VA, December 15, 1994, p. 20, Figure 2.

tional capability would be available that would consist of ten conventional HAE UAVs (two engineering and eight field models), all with payloads; three CGSs (one engineering and two field demonstration models); two LO HAE UAV engineering models; and up to eight LO field demonstration models. Additionally, there would be 200 trained personnel (military and contractor).[2] DARO, the sponsoring agency, would fund the program through the completion of Phase III. The Air Force, as the lead agency, was expected to program operations and support (O&S) funds for the maintenance and operation of the residual systems.

This early management plan did not specify the number of flight hours expected from either Phase II or Phase III. It did, however, make clear the importance of the flight test program in the context of the ACTD construct (a full three years was planned, representing more than half the ACTD schedule). The plan also introduced the notion of contractor responsibility for the flight test program as well as early user participation through the field test portion of the flight test program.

The initial Master Test Plan for Global Hawk, developed by the air vehicle contractor, Teledyne Ryan Aeronautical (Ryan), was released on March 30, 1995, during Phase I of the program. A revised test plan was issued in November 1995, early in Phase II.[3] This document, which incorporated IPT comments and formed the basis of the Phase II flight test program, laid the groundwork for the roles and responsibilities of government and contractor while also setting the procedures for the conduct of flight tests. The government was invited to participate at all IPT levels, but the contractor was to be in charge. All tests would follow a previously approved test plan ranging in scope from engineering design validation to complex flight tests requiring interagency (FAA, test range) coordination.

The primary objective of Phase II testing was to measure system technical performance against the characteristics of the contractor's written system specification. Technical performance measures

---

[2]Op. cit., p. 20. Assumes no loss of assets during the test program.

[3]Teledyne Ryan Aeronautical, *Master Test Plan for the Tier II Plus High Altitude Endurance (HAE) Unmanned Air Vehicle*, San Diego, CA, Report No. TRA-367-5000-67-R-001A, November 17, 1995.

(TPMs) and system maturity matrix (SMM) goals were identified and cross-referenced to the preliminary system specification. The TPMs were related to the technical and payload incentives embodied in Attachment 5 of the Ryan Agreement. The test plan included ground-based system and subsystem tests required to demonstrate flight readiness. Up to seven airworthiness flights and up to nine payload flights were planned. For each flight, both primary and secondary objectives were identified.

The small number of flights in the Global Hawk test plan (16) was facilitated by the relatively small area of the flight envelope inherent in HAE aircraft. The elapsed time/vehicle speed/altitude mission profile for every test flight would be similar with the exception of time at cruise. The primary stress on the air vehicle and sensor payload derives from increasing dwell times at high altitude. Thus, endurance and sensor performance are tested at various altitudes.

The December 30, 1997, HAE UAV ACTD management plan (version 7) gave somewhat more attention to flight test planning.[4] Phase II engineering tests conducted separately for both Global Hawk and DarkStar were to begin in the middle of FY 1997 and were to last through the end of FY 1998. The Air Force Flight Test Center (AFFTC) was to handle airspace coordination for Global Hawk. Phase II activities called out in the plan included:

- Defining user segments to facilitate early developer and user integration;
- Establishing system baselines and characterization;
- Developing planning tools and employment techniques for the user; and
- Publishing the HAE UAV ACTD assessment operations plan (OPLAN), which establishes user military utility assessment (MUA) methods and objectives.

---

[4]Note that all management plans through the December 1997 version were written by the DARPA joint program office. The last (version 7) was the one that transitioned to the Air Force; no update was approved prior to the October 1998 management transition.

Global Hawk engineering development tests would include seven airworthiness flights and nine payload flights over a 12-month period. The only listed criterion for transition from Phase II to Phase III was approval from the HAE UAV Oversight Group.

Phase III D&E testing was to begin in the last quarter of FY 1998 and was to continue through the end of FY 1999. Exercises were to be identified by asking the commanders in chief (CINCs) for their joint mission-essential task lists (JMETLs), whose purpose was to define tasks in which HAE UAV participation could make a difference. The number of exercises would be chosen on the basis of the number needed to sufficiently characterize the system's utility. Phase III was intended to be a combined testing of DarkStar and Global Hawk, moving from scripted demonstrations to more complex participation in scheduled exercises. Contractor maintenance and operational support was to be used. The United States Atlantic Command (USACOM) was to plan and execute the Phase III demonstrations. The results were intended to inform the then-planned force mix and MUA decisions at the completion of the ACTD.

By December 1997, cost increases in the program had reduced the number of available vehicles for Phase III D&E. The management plan stated that up to eight Global Hawk and/or DarkStar air vehicles and two CGSs would support user demonstrations.

Planning for Phase III D&E testing was also initiated in 1995. The ACTD office in USACOM stood up in February 1996 with roughly ten programs, one of which was an HAE UAV. USACOM began formally planning for the HAE UAV MUA in May 1996. Toward the end of Phase III, the Joint Forces Command (JFCOM) had six people working the program, but most of the work was done by two individuals, one of whom was the designated operational manager (OM) in the HAE UAV ACTD program structure.[5]

The program office defined an MUA algorithm in late 1995. The algorithm was presented in charts at the monthly Home Day meeting in January 1996. Nominal values were calculated at design reviews in support of the algorithm. The MUA algorithm shown below results in a single-point estimate.

---

[5]Colonel John Wellman, U.S. Air Force, and Walt Harris.

> MU = f(area coverage × no. of targets/day × no. of UAVs required × targeting accuracy × timeliness × image quality × flexibility × (threat, CONOPS, mission planning, survivability suite).

Either JFCOM rejected the algorithmic approach or both the joint program office (JPO) and JFCOM had abandoned it by the time detailed planning for D&E began. As a result, the algorithm appears to have played no role in the assessment. In any case, a single-point multidimensional quantitative estimate is not an appropriate indicator of military utility for a system as complex and capable as the HAE UAV. However, the algorithm does suggest the types of metrics that could be used in the MUA, although their relative importance was not indicated.

USACOM/JFCOM originally hired the Defense Evaluation Support Activity (DESA) to help design and execute the D&E phase. Shortly thereafter, DESA became Detachment 1 of the Air Force Operational Test and Evaluation Center (AFOTEC) as a result of a reorganization. JFCOM was somewhat concerned that this organizational change would result in the incorporation of Air Force biases, but Detachment 1 remained a separate organization within AFOTEC and maintained its objectivity. In collecting and organizing test data, Detachment 1 did use many of the same formats and styles as the rest of AFOTEC. Detachment 1 would collect the data and write the quick-look reports and summary reports. JFCOM would write the final MUA defining the HAE UAV's value added to the D&E exercises. The JFCOM CINC had final approval authority for the MUA. This CINC was involved in and supportive of the HAE UAV ACTD program. Beyond the specific approach laid out in the integrated assessment plan (IAP), he set up a strict criterion for determining military utility: to *prove* that the HAE UAV makes a difference in operations.

JFCOM/AFOTEC initially stated that approximately 2000 flight hours would be needed to demonstrate military utility. At the expected flight-hour accumulation rate, this roughly translates into a three-year flight test program. The number of flight hours required was driven by operational suitability concerns. Given the shortened D&E in which an unknown number of flight hours would be accumulated, a "crawl-walk-jog-run" pattern was eventually agreed on in which the pace of flight test activity was to increase over time.

The D&E phase was designed from its inception to collect as much traditional development test/operational test (DT/OT) data as possible to facilitate a tailored and streamlined post-ACTD development program. In early 1996, when the D&E plan was first put together, JFCOM asked AFOTEC to identify the type of information needed in a traditional DT/OT flight test program as well as to estimate approximately how many flight hours that program has traditionally taken. The D&E phase was structured on the basis of that input, recognizing that all such information could not be collected. The plan included 1200 hours of flight test, derived from Air Force "maturity" statistics. Establishing the system's reliability, maintainability, and supportability characteristics constituted the primary driver.

The November 1995 Master Test Plan for Global Hawk also briefly discussed a notional Phase III field demonstration program. It noted the participation of the Tier III– and added that system capabilities should be expected to evolve as the results of each flight test are used to enhance system performance. The assessment of effectiveness and suitability was identified as the primary Phase III objective and criterion for military utility. Effectiveness was defined as including long operational range, extended endurance, interface with existing command, control, communications, computers, and intelligence (C4I) architecture, high-resolution sensors, and geolocation accuracy. Suitability was said to include flexibility in basing, transportability, safety, availability, maintainability, and supportability. The last three criteria are particularly notable for an ACTD. Phase II assets may be used in support of demonstration activities as needed. The stated objective was to demonstrate military utility to the user while simultaneously demonstrating system performance characteristics. Demonstration tests would be formally planned with mission cards, go/no-go criteria, support plans, and the like. The notional beginning of field demonstrations was January 1998. The initial pace was four 24-hour flights per month, increasing to 20 flights per month by the sixth month. Operations were expected to transition from a "training condition" to a "deployable condition." The simultaneous deployment of a full system that included four Tier II+ vehicles, two Tier III– vehicles, one ground segment, and a spares kit, together with continued operations of 250 flight hours per month from the main operating base (MOB), was envisioned.

After the completion of the 16 engineering flight tests, cumulative flight hours were expected to rapidly increase as additional assets became available (e.g., eight additional Tier II+ air vehicles and two additional CGSs delivered during the 24-month Phase III). Table 2.1 shows the expected buildup of flight hours for the Tier II+ by quarter. By the end of 1998, eight UAVs were expected to be flying; all ten were expected to be flying by the end of the first quarter of 1999 (representing a delivery rate of two air vehicles per quarter beginning in the second quarter of 1998).

The DARPA JPO's review of the contractor's initial test plan was somewhat critical of that plan in terms of the detail made available, the lack of a master test matrix, and the relationship of Simulation and Integration Laboratory (SIL)/nonflight tests to the flight test program.[6] It would appear that these problems were adequately addressed in subsequent plans.

The test plan states the expectation that live fire testing will not be required for the HAE UAV ACTD program based on a September 19, 1995, letter from the Deputy Director of Live Fire Testing.

Table 2.1

Early Notional Phase III Field Demonstration Plan

	1998 (quarter)				1999 (quarter)			
	1st	2nd	3rd	4th	1st	2nd	3rd	4th
Flights per quarter	12	24	60	60	60	60	60	60
Flight hours per quarter	300	600	1500	1500	1500	1500	1500	1500
Cumulative flight hours (Phase III)	300	900	2400	3900	5400	6900	8400	9900

SOURCE: Teledyne Ryan Aeronautical, *Master Test Plan for the Tier II Plus High Altitude Endurance (HAE) Unmanned Aerial Vehicle,* San Diego, CA, Report No. TRA-367-5000-67-R-001A, November 17, 1995.

---

[6]See Home Day charts, August 1996.

In general, the HAE UAV ACTD test plan appeared similar to that of a traditional program in its use of formal test plans, performance matrices, and readiness reviews and in its linking of ground and flight testing. The key difference lay in the fact that test planning and execution were to be led by the contractor. With only 16 flights planned, the Phase II engineering test plan seemed rather thin, but this is understandable given the limited flight envelope of HAE aircraft. By contrast, the Phase III D&E phase laid out early in the program seemed somewhat ambitious. Overall, the plan appeared to be unrealistic, especially given the technical unknowns associated with a system that represented such a fundamental departure from prior experience.

FLIGHT TEST PROGRAM EXECUTION

The first air vehicle to fly in the HAE UAV ACTD program was DarkStar, whose first flight took place on March 29, 1996. This air vehicle was destroyed on April 22, 1996, as it was taking off for its second flight.[7] The HAE UAV ACTD flight test program resumed activity on February 28, 1998, some 22 months later, with the first flight of Global Hawk. DarkStar resumed flights with air vehicle 2 on June 29, 1998, some 26 months after the loss of the first DarkStar. The last formal ACTD flight test, conducted on July 19, 2000, consisted of a functional checkout of Global Hawk air vehicle 5.

Excluding the first two DarkStar flights, the flight test program for both air vehicles appears to have been fairly cautious. A pattern of learning—test, analyze, fix, test—is clearly evident for both Global Hawk and DarkStar.

The flight test program also represented a steady progression in learning, particularly for Global Hawk. Each flight generated more confidence in both air vehicle and ground systems. Minor anomalies occurred but were resolved. Overall, the flight test program was characterized by a steady, planned envelope expansion and capability demonstration for both engineering and D&E elements of the

---

[7]See Drezner, Sommer, and Leonard, *Innovative Management in the DARPA High Altitude Endurance Unmanned Aerial Vehicle Program*, 1999, for more detail on the crash, its causes, and its consequences.

flight test. Flash reports and quick-look reports[8] were written for all flights; these constituted after-action reports for each sortie and supported a continuous MUA process. The program had no formal performance requirements, but information continually emerging from the flight test program was intended to support operational requirements document (ORD) development.

Summary information on the ACTD flight test program is provided in Table 2.2. Not surprisingly, neither DarkStar's nor Global Hawk's flight test program was executed as originally planned.

DarkStar

The first flight of DarkStar air vehicle 1 was characterized by significant anomalies. These were not sufficiently resolved by the time of the second flight on April 22, 1996, resulting in the vehicle's destruction on takeoff.

DarkStar air vehicle 2 first flew 26 months later. As Figure 2.1 shows, DarkStar flight testing with air vehicle 2 lasted for six months, a period in which only six hours of flight time were accumulated.

Table 2.2

Summary Flight Information for ACTD Air Vehicles[a]

Air Vehicle	First Flight Date	Total Cumulative Flight Hours
DarkStar 1	March 29, 1996	0.75
DarkStar 2	June 29, 1998	6.0
DarkStar 3	n.a.	n.a.
Global Hawk 1	February 28, 1998	328.3
Global Hawk 2	November 20, 1998	55.1
Global Hawk 3	August 12, 1999	121.8
Global Hawk 4	March 24, 2000	167.8
Global Hawk 5	June 30, 2000	64.1

[a]Data as of September 14, 2000.

[8]Flash reports were brief summaries written and distributed immediately following each sortie. Quick-look reports provided more detailed descriptions of each flight test, assessed the extent to which flight test objectives had been achieved, and documented possible problems and anomalies.

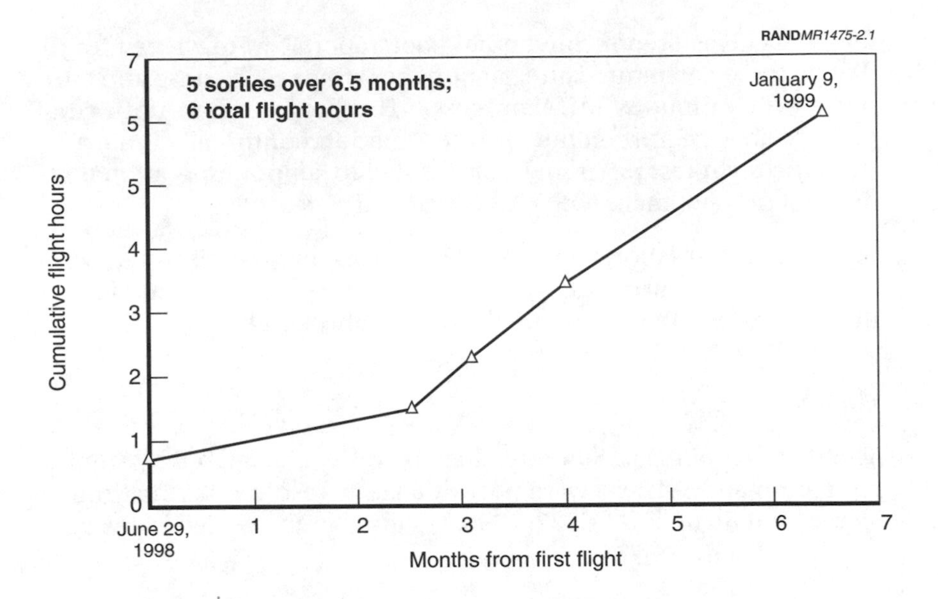

Figure 2.1—DarkStar Air Vehicle 2 ACTD Flight Test Program

DarkStar was terminated a day before its planned seventh flight and never entered Phase III D&E. An important concern is whether the shortened flight test program yielded useful results in the form of either technical characteristics relating to air vehicle configuration or lessons regarding operating procedures.

Global Hawk

The first flight of Global Hawk occurred on February 28, 1998 (Figure 2.2). Phase II engineering flight testing included the first 21 sorties. Phase III flight testing included sorties 22 through 58.[9] The Phase III flight test program included D&E sorties associated with a specific military exercise as well as some additional functional checkout and follow-on engineering flights. Post–Phase III flight testing, which in-

[9]The flash report for Flight 58, air vehicle 5, sortie 4 specifically states that this was the final Phase III sortie.

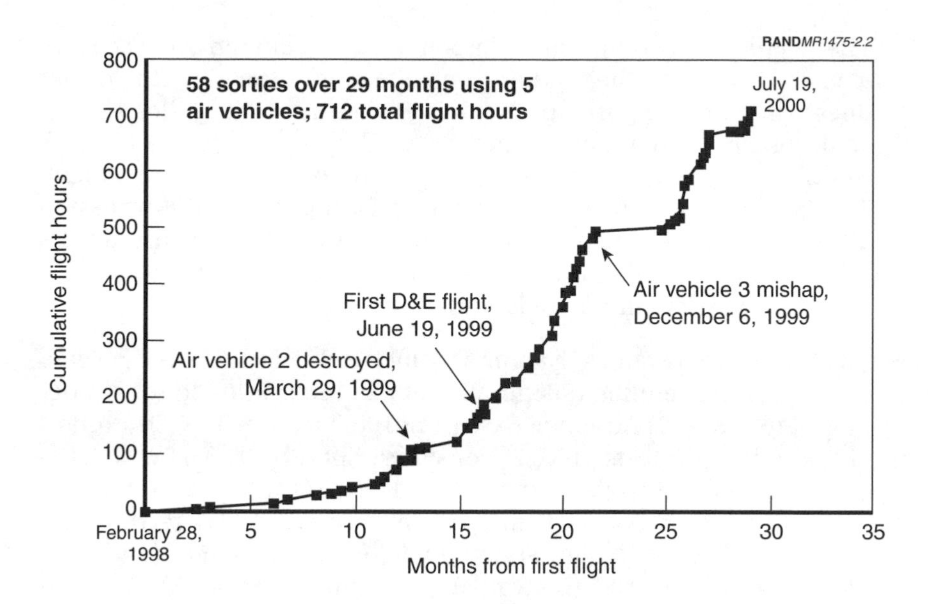

Figure 2.2—Global Hawk ACTD Flight Test Program

cluded Flight 59 and subsequent (not included in Figure 2.2), supported both preparation for the Australian deployment and another D&E exercise (Joint Expeditionary Forces Experiment [JEFX]). No more than two air vehicles participated in the ACTD flight test program at any one time because air vehicle 2 was destroyed several months before air vehicle 3 became operational and because the air vehicle 3 postflight taxi mishap occurred prior to the first flight of air vehicle 4. Global Hawk's 29-month ACTD test program reflects a clear pattern of building confidence in the system; the pace of flight testing increased significantly as the program proceeded.

Common Ground Segment

Both DarkStar and Global Hawk were designed with distinct ground segments consisting of a launch element and a mission element. In early 1995, planning for the CGS was initiated. The CGS essentially incorporated DarkStar functionality into the Global Hawk ground segment.

The CGS[10] had two primary components: the launch and recovery element (LRE) and the mission control element (MCE). The former does what one might expect given its name: it handles the air vehicle's taxi, takeoff, and climb-out at the beginning of a flight as well as its descent, approach, landing, and taxi at the end of a flight. The latter takes control of the aircraft during its climb to mission altitude; controls the aircraft and tasks the sensors through its mission; and flies the aircraft toward its landing site until the LRE takes over in preparation for landing.

MCE 1 was delivered to Ryan on October 1997. LRE 1 was delivered to Ryan on November 1996 for air vehicle integration, and was deployed to Edwards Air Force Base (EAFB) in October 1997 in support of Phase II flight tests. LRE 2 (considered part of CGS 1) was delivered to Boeing in November 1998 for DarkStar integration but was returned to Raytheon in February 1999, where it was retrofit to a Global Hawk–only configuration and delivered to Ryan in June 1999. MCE 2 was delivered to Ryan in September 1999. MCE 1 was returned to Raytheon in December 1999 for an upgrade and was then returned to Ryan in March 2000. LRE 3 (which was considered part of CGS 2) was delivered to Ryan in November 1999. LRE 1 was returned to Raytheon for an upgrade in February 2000 and was returned in June 2000. All LREs and MCEs supported Phase III testing at some point.

CGS performance was clearly a major concern during flight test, as much of the HAE UAV's capability was embodied in the ground segment. The performance of both the LRE and the MCE appeared satisfactory during both Phase II and Phase III testing; we uncovered no evidence of significant performance problems. However, ground segment–related issues pertaining to operational procedures, to the roles and responsibilities of functions within the CGS system, and to the time-consuming and difficult mission planning process did arise during test. Operational issues were resolved as more experience was gained with the system. The mission planning system is the target of major improvements in post-ACTD plans.

---

[10]Now simply known as the ground segment.

The CGS was tested in its Global Hawk–only configuration but did not play a role in the shortened DarkStar flight test program, as DarkStar used its own uniquely developed ground segment. The CGS was never fully tested during Global Hawk's flight test program. Following DarkStar's cancellation, there was no opportunity to determine whether the "common" aspect of the CGS functioned adequately. Owing to the lack of trained personnel and spares that allowed just one air vehicle in flight at any one time, the CGS never demonstrated the ability to control multiple air vehicles simultaneously.

POLICY IMPLICATIONS

From an acquisition policy perspective, there are three important aspects of the HAE UAV ACTD flight test program:

1. Contractors were given significantly increased responsibility and had the lead for flight test planning and execution.
2. Development testing and operational testing were concurrent during the ACTD.
3. The program was characterized by early user involvement in planning and executing operational demonstrations prior to the completion of development testing.

The application of Section 845 OTA and the delegation of design and management authority to the contractors resulted in the contractors' designation as the lead for test planning and execution. In traditional programs, the flight test center associated with the test range is usually designated the responsible test organization (RTO) for both technical and safety elements. In this case, the RTO's responsibilities were shared between the contractors, the program office, and AFFTC. This increased contractor involvement changed the institutional dynamic of the flight test program. Some participants claim that these changes resulted in a more targeted and flexible flight test program with faster decisionmaking and approval. Other participants, however, suggest that giving the contractor the execution lead contributed at least indirectly to the various mishaps that affected the program. All participants agree that contractors generally lack the operational perspective to successfully and safely run a flight test program and thus require support from on-site operators and from the test center.

The ACTD user (USACOM/JFCOM) had significant input into flight test planning and execution, particularly in the D&E portion. Without JFCOM's involvement, supporting exercises would have been administratively impossible for the JPO to accomplish as the JPO had neither the authority nor the experience to define and manage its participation in joint military exercises. Representatives of the force provider—i.e., the 31st Test and Evaluation Squadron (TES) from the Air Combat Command (ACC)—were major participants in the flight test program, although they were not included in the plan. Such user participation in early testing, although not common in traditional acquisition approaches, benefited all participants in the HAE UAV ACTD program. In addition, Phase III D&E included combined DT/OT on almost every flight. This is highly unusual early in a program.

Two key acquisition policy–related issues arise with respect to the Global Hawk system flight test program. The first lies in determining whether the flight test program was sufficient to satisfy the objectives of the ACTD. Put another way, did the flight test program generate sufficient information to adequately characterize the system's effectiveness, suitability, and interoperability? Flight testing is somewhat different today than in the past. The measure of merit for a program no longer consists of the number of sorties generated or cumulative flight hours; instead, it pivots on whether adequate information was collected. Considerably more information is collected during each sortie today than in the past.[11]

Second, to what extent can the accomplishments of the ACTD flight test program be used to tailor and shorten post-ACTD flight test activities? In particular, to what extent can Global Hawk's ACTD flight test experience satisfy developmental and operational testing requirements mandated for engineering and manufacturing development (EMD) programs?

The remainder of this document addresses these questions.

---

[11]See William B. Scott, "F-22 Flight Tests Paced by Aircraft Availability," *Aviation Week & Space Technology*, October 16, 2000, pp. 53–54.

Chapter Three

GLOBAL HAWK

PHASE II ENGINEERING DEVELOPMENT

According to the earliest program documentation, Global Hawk flight testing was scheduled to begin in December 1996.[1] As of September 1997, however, the integrated mission management computer (IMMC) software was not yet ready for taxi or flight. Such problems were tracked by the DARPA JPO at its monthly Home Day meetings and were documented in the briefing charts supporting those meetings. As previously discussed, software development and integration problems contributed significantly to the 14-month delay of the first flight, which occurred on February 28, 1998.[2]

The pace of the flight test program was largely determined by the fact that Global Hawk is both a large, complex airplane and an autonomous UAV. For the most part, the character of the system determined the design and execution of the engineering portion of the flight test program.

Development testing constituted the first 21 flights, the last three of which (flights 10–12 for air vehicle 1) were considered follow-on

---

[1]See Drezner, Sommer, and Leonard, *Innovative Management in the DARPA High Altitude Endurance Unmanned Aerial Vehicle Program,* Table 3.3, 1999, p. 55. DARPA's *HAE UAV ACTD Management Plan,* December 15, 1994, and Teledyne Ryan Aeronautical's *Master Test Plan,* November 17, 1995, indicated a planned first flight in January 1997, a one-month difference.

[2]See Drezner, Sommer, and Leonard, *Innovative Management in the DARPA High Altitude Endurance Unmanned Aerial Vehicle Program,* 1999.

DT.[3] Flight 13 of air vehicle 1, conducted on June 19, 1999, was the first D&E flight test. The Phase II engineering development flights roughly accomplished what a traditional demonstration/validation program would. All subsequent flights during Phase III were dedicated to demonstrating the operational system, performing functional checkout of additional air vehicles, or testing configuration improvements. The demonstration flights are unique to the ACTD process. The configuration change test and functional checkout flights are EMD-type activities.

Figure 3.1 shows the gradual buildup of flight hours during Phase II testing. Air vehicles 1 and 2 flew a combined total of 21 sorties over a 16-month period, accumulating 158 flight hours. Air vehicle 1 flew 12 sorties for 103 flight hours; air vehicle 2 flew nine sorties, accumulating 55 flight hours. Flight 11 of air vehicle 1, the 20th sortie of the program, was the longest, lasting 27.2 hours. Air vehicle 1 mainly flew airworthiness sorties, while air vehicle 2 primarily flew payload checkout sorties.

The first aircraft explored the entire flight envelope in five flights. Each additional aircraft required approximately three productive flights to be completely checked out in the envelope. The only true expansion of the flight envelope occurred when the elapsed time was extended at higher altitudes, as the possibility of extremely cold temperatures at such altitudes had the potential to stress the system. Flight test personnel at EAFB commented that it took them nine sorties using air vehiicle 1 to get comfortable with the system.

All Phase II flights were either airworthiness (11) or payload (10) functional checkout flights. Objectives were not met in four of the 21 flights (one airworthiness and three payload). Objectives were at least partially met in the other 17 sorties. In cases where objectives were not met, the mission was generally reflown. If only one or two objectives from the list for a specific flight were not met, these were often added to the next mission.

---

[3]Although the program had planned to fly the initial flights at a faster rate, it ended up under budget owing to a slower-than-planned increase in pace and to delayed hiring of additional test personnel.

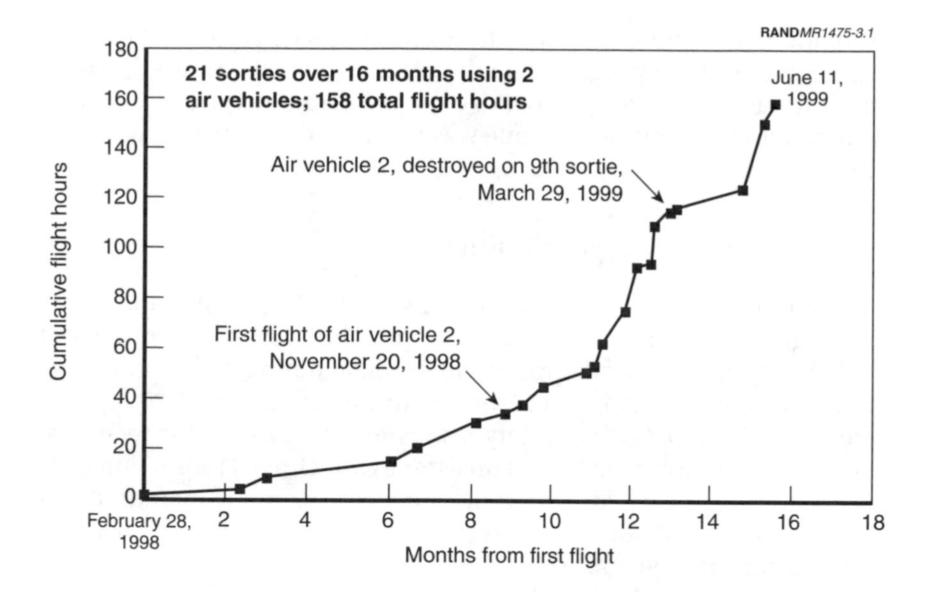

Figure 3.1—Phase II Global Hawk Flight Test Program

Engineering development tests for Global Hawk occurred in both Phase II and Phase III of the ACTD. Development tests culminated in certification that Global Hawk was ready for operational testing; this was called the Phase II assurances process. Phase II assurances provided the HAE UAV Oversight Council sufficient information to approve entry into the next phase.[4]

The Phase II flight test program did slip by several months. The original plan called for a 12-month program; the actual program was 16 months. Phase II was to be completed when program management transitioned from DARPA to the Air Force, which actually took place in October 1998. The phase was actually completed some nine months later, when the first Phase III D&E flight took place in June 1999. Phase II testing did accomplish the majority of its objectives, as it allowed for an initial characterization of both air vehicle and synthetic aperture radar (SAR) performance. However, the program

---

[4]Command, Control, Communications, Computers, and Intelligence Support Plan (C4ISP) final report, November 15, 2000.

did not demonstrate the ability to control two Tier II+ air vehicles at a time. Nor did the program test the electro-optical/infrared (EO/IR) sensor sufficiently to characterize its performance; the only EO/IR subsystem available at that time was lost in the destruction of air vehicle 2.

Contractor Test Responsibilities

The contractors (Lockheed Martin Skunk Works [LMSW] and Ryan) were given significantly increased responsibilities in the conduct of the flight test program. This included designating the test director and taking the lead for test planning and execution, with assistance from AFFTC at EAFB for safety issues and from ACC for technical support. This arrangement is consistent with Other Transaction (OT) implementation in this program. The arrangement increased demands on contractor flight test personnel while reducing them on flight test center personnel.

The original flight test director at Ryan had no experience with large aircraft programs. However, the current Ryan flight test director at EAFB, who was hired in January 1997, played a critical role in shaping the Global Hawk flight test program and its execution. On the basis of his experience in the Air Force on the Lightweight Fighter (LWF) program,[5] he involved user representatives—the 31st TES from ACC—in all aspects of flight testing in efforts to introduce an operational flavor.[6] He did this over the objection of the Ryan program manager at the time. In the end, Ryan, the government program office, and the 31st TES collaborated to get the 31st more involved.

The contractor's extensive test responsibilities appear to have had little substantive effect on the Phase II engineering development flight test program. This is not surprising given that contractor flight testing is the norm in early engineering development. Decisionmaking was perhaps a little faster, but interagency coordi-

---

[5]The LWF program was one of several streamlined competitive prototyping efforts run by the Air Force in the early 1970s. This program, which included the YF-16 and YF-17, is considered particularly successful in that both prototypes led to new operational systems.

[6]The LWF did this by including operational pilots.

nation was somewhat more difficult. In any case, government agencies (the Global Hawk System Program Office [GHSPO] and AFFTC) played significant roles in the execution of the flight test program, and the SPO supported Ryan in its test planning responsibility.

Destruction of Air Vehicle 2

The major setback during Phase II flight testing was the destruction of air vehicle 2 on March 29, 1999, during the program's 18th sortie. The loss of air vehicle 2 and its payload was estimated at $45 million. Of more importance, however, was the fact that the program lost its only integrated sensor suite. The crash was due to a lack of proper frequency coordination between the Nellis Air Force Base and EAFB flight test ranges. Essentially, Nellis officials who were testing systems in preparation for Global Hawk's first planned D&E exercise were unaware that Global Hawk was flying over China Lake Naval Air Weapons Station, which is within EAFB's area of responsibility. When Nellis tested the flight termination code, Global Hawk responded exactly as designed.

Air Force frequency management procedures were not designed to accommodate an autonomous high-altitude UAV. High-altitude flight creates a much greater distance for receiving line-of-sight commands. The absence of a human in the loop, either onboard the aircraft or on the ground, did not permit the unintended flight termination command to be disregarded. The procedures that allowed these circumstances to arise were thus an Air Force–wide problem rather than one specific to Global Hawk. These procedures were changed as a result of this accident, thereby precluding a similar incident in the future.[7]

One circumstance leading to the destruction of air vehicle 2 was the contractor's decision to refly on Monday the sortie that had been aborted three days before. Some participants believe that the flexibility to execute the reflight so quickly stemmed from the contractor's status as lead for test program execution. Others believe that

---

[7]Excerpted from the Accident Investigation Board report released on December 22, 2000. See "Poor Communications Management cited in Global Hawk UAV Crash," *Inside the Air Force*, Vol. 11, No. 1, January 7, 2000, pp. 9–10.

the Air Force would have made the same decision and executed the same quick turnaround. Most participants stated that the destruction of air vehicle 2 was not a result of contractor involvement in the test program because Ryan relied on AFFTC for test support in any case. However, the incident report states that Ryan did not follow established notification procedures for the revised mission. Some participants further noted that Ryan had in fact followed these procedures and had provided the necessary information to the appropriate office at EAFB. Unfortunately, the person who normally handles frequency management coordination at EAFB was on leave that day. Other participants noted that had AFFTC been the RTO, it might not have approved the Saturday workload that was required to support a Monday flight owing to manning and flight operational tempo issues. Considering all of these views, it is not clear if the contractor's designation as lead for test program execution played a role in the loss of air vehicle 2.

PHASE III DEMONSTRATION AND EVALUATION

Global Hawk progressed into the Phase III user D&E phase in June 1999. DarkStar was terminated in January 1999, well before the completion of its engineering flight tests or the start of the D&E period.

Phase III D&E focused on generating information to support the MUA. Although the bulk of the Phase III effort was devoted to planning and executing D&E exercises, some flights supported engineering testing for vehicle functional checkout and acceptance, sensor checkout, and wing pressure validation. Other sorties supported both follow-on engineering development and D&E exercises, as tests were performed as the air vehicle transited to its position in support of a given D&E exercise.

The D&E IPT operations plan dated September 1997 documents the MUA process envisioned at that time. Both USACOM and ACC participated on the IPT and in the development of the plan, with USACOM acting as the chair and ACC providing personnel and some post-ACTD operations planning. Data taken from the D&E flights were intended to serve three purposes: (1) to inform the MUA; (2) to support the post-ACTD decision process; and (3) to characterize the ACTD configuration capability for use of the residual assets.

Effectiveness, suitability, and interoperability were the three top-level operational parameters considered in the MUA. From these, operational issues were derived, each of which had several subobjectives and associated metrics. An assessment plan was developed for each exercise, and the results were documented in an after-action report. The results documented in each after-action report were combined into the overall assessment. The entire process, including the details of objectives and subobjectives, data collection methods, and recommended training for data collection personnel, was documented in the Integrated Assessment Plan (IAP) dated June 1998.

Given the complexity and novelty of this undertaking, the MUA process described in these later documents appeared reasonable. However, there were some noticeable gaps. Absent from the IAP, for example, was a more precise description of how the information generated during the exercises would be used in the MUA. Also omitted was information on the relative importance of the objectives and subobjectives of the MUA determination and details on how the information gained would support requirements generation and post-ACTD planning.

When Phase III D&E was shortened as a result of schedule slips in other aspects of the program, USACOM/JFCOM expressed concern regarding the adequacy of the information that would be generated by the diminished number of exercises and flight hours. It was feared that the abbreviated flight test program would not support a definitive MUA.

In the shortened Phase III, quick-look reports were produced after each demonstration exercise. These reports, written by AFOTEC Detachment 1, documented key experiences during each demonstration and noted key problems. The information collected at each exercise differed depending on the objectives laid out in the assessment plan for that exercise. In general, the quick-look reports were carefully done and consistent both across demonstration exercises and with IAP criteria and procedures. These documents were synthesized in a series of sequenced, cumulative reports entitled "crawl," "walk," and "run"/final, forming the basis of the final MUA document. The process for aggregating and communicating the results of the D&E was revised from the original plan to this approach, providing incremental reports.

Figure 3.2 shows cumulative flight hours for the Global Hawk Phase III flight test program. Over a 13-month period, 37 sorties were flown, yielding a total of 554 flight hours. Four air vehicles were used, never more than one at a time. The LRE and the MCE used in Phase II were supplemented by two additional LREs and one additional MCE during Phase III in a Global Hawk–only configuration.[8]

D&E missions accounted for 21 sorties supporting 11 exercises that totaled 381 flight hours. The exercises in which Global Hawk participated were as follows:

- Roving Sands 1 (June 19, 1999)
- Roving Sands 2 (June 26, 1999)
- Roving Sands 2B (June 27, 1999)

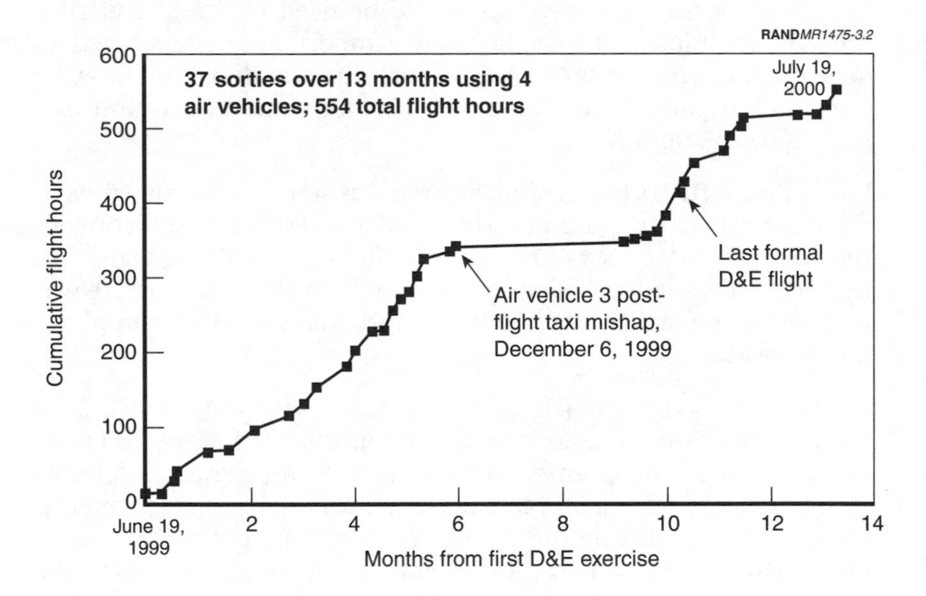

Figure 3.2—Global Hawk Phase III Flight Test Program

[8]MCE 2 and LRE 2 were already configured in the CGS design that incorporated DarkStar functionality. They were in various stages of final integration when DarkStar was terminated. Both units were returned to Raytheon for removal of DarkStar functionality and for subsequent redesign into a Global Hawk–only configuration.

- Extended Range 1-1 (July 15, 1999)
- Extended Range 1-2 (July 27, 1999)
- Extended Range 2/JEFX/Combined Arms Exercise (CAX) (August 30, 1999)
- CAX 99-10 U.S. Marine Corps Exercise (September 9, 1999)
- Extended Range U.S. Navy Seals (October 4, 1999)
- Extended Range 3-02 U.S. Navy Seals and close air support (CAS) (October 8, 1999)
- Extended Range 4-01 Alaska (October 19, 1999)
- Extended Range 4-02 Alaska (October 25, 1999)
- Desert Lightning II (November 9, 1999)
- Desert Lightning II (November 13, 1999)
- Desert Lightning II (November 17, 1999)
- Joint Task Force (JTF)-6 Sortie 1 (December 3, 1999)
- JTF-6 Sortie 2 (December 6, 1999)
- D&E Deployment to Eglin Air Force Base (April 20, 2000)
- Linked Seas–1 (May 8, 2000)
- Linked Seas–2 (May 11, 2000)
- Joint Task Force Exercise (JTFEX)00-1 (May 18, 2000)
- JTFEX00-2 (May 19, 2000)

A number of these D&E flights represented reflights of unsuccessful or partially successful prior sorties. Roving Sands 2B was a reflight of the previous mission, which was curtailed when the SAR would not come online. The third Desert Lightning II flight was a refly of the previous mission, which was curtailed owing to an IMMC failure. The JTF-6 Sortie 2 mission was curtailed, and a safe landing was accomplished using a contingency flight and landing profile, after which the Air Vehicle 3 taxi accident occurred as the aircraft was preparing to taxi off the runway. This accident led to a stand-down of all flight testing from December 1999 to March 2000.

The first four D&E exercises (Roving Sands 1999, Extended Range 1 and 2, CAX 1999) comprised the crawl phase. The next five (Extended Range 3 and 4, JTFEX, National Training Center [NTC], Extended Range/CJTR) comprised the walk phase. The run (final) phase included the remaining exercises.

Extended Range 4-01 to Alaska was the first flight outside continental U.S. (CONUS) airspace. The flight after the stand-down was the deployment to Eglin Air Force Base, Florida, which included extensive flight time in Federal Aviation Administration (FAA)-controlled airspace. Some program participants considered the deployment to Eglin Air Force Base the most critical D&E flight test. The East Coast deployment included both ground and air vehicle deployment; several exercises over the East Coast and the first Linked Seas mission included a transatlantic flight to Portugal.

Thirteen sorties and 152.3 flight hours conducted during Phase III were for the functional checkout of air vehicle 3 (four flights), air vehicle 4 (five flights), and air vehicle 5 (four flights). An additional three sorties totaling 20.7 hours were conducted for other engineering development objectives, such as wing pressure validation.

Of the 37 missions flown in Phase III, 26 fully met their objectives, six did not meet their objectives, and five met their objectives to some degree. The D&E program concluded in May 2000 after JTFEX00. Air vehicle 4 was redeployed to EAFB from Eglin Air Force Base on June 19, 2000.[9] The next four flights were air vehicle 5's functional checkout. Flight 4 of air vehicle 5 took place on July 19, 2000, and was the last sortie in Phase III of the ACTD program.

The quick-look reports noted that many of the problems found during testing were procedural in nature, reflecting the process of learning how to operate the system and integrate it into military operations. As would be expected, performance—specifically the timeliness of imagery transmission—improved from the crawl to the walk stage, reflecting learning and updated procedures.

[9]See the June 2000 monthly acquisition report (MAR) from the GHSPO.

The 31st Test and Evaluation Squadron

The 31st TES is a field operations unit for ACC. The ACC Director of Operations (ACC/DO) directed that the unit stand up at EAFB in the fall of 1997. The manning of the 31st TES grew to 16 to 20 full-time personnel. The 31st TES was not part of the program plan, and there was no formal process for implementing a relationship between Ryan and the 31st TES. However, Section 845 OTA was flexible enough to allow such interactions.[10] ACC understood the importance of its input to the success of the Global Hawk test program and therefore paid for 31st TES personnel by using manpower slots from the Predator program budget. The 31st TES became a core partner in the flight test program, at times providing 50 percent of mission planning capabilities and 50 percent of operations. The 31st TES added significant value in terms of configuration changes, mission planning, command-and-control officer (CCO) experience and technical orders, and an operational perspective. The unit helped shape system capabilities and operational procedures.

To address targets of opportunity, the 31st TES developed the capability for in-flight dynamic sensor retasking during operational demonstrations. Essentially, unit operators learned to "trick" the aircraft to image targets that were not part of the mission plan. As a result of the 31st TES's desire to make the system more operationally useful, its operators also learned to use the vehicle in a manner that differed from its intended design. Global Hawk was designed to be fully autonomous and, according to one operator, was expected to be used like a "wind-up toy." With the influence of the 31st TES, the system proved to be flexible and interactive.

In general, the 31st TES provided both military experience and an operational perspective and, as a result, became the Air Force expert on HAE UAV operations, supporting briefings to senior Air Force officials. The 31st TES was clearly an asset to the program and was involved to an extent more than is usually seen in traditional flight test

---

[10] Agreement modifications between Ryan and the GHSPO were strictly negotiated because of the Ryan program manager's insistence on having everything precisely specified. No allowance for 31st TES participation was ever put in writing. The Ryan flight test director gave the 31st TES a role despite the disapproval of his own management.

programs. From its operational perspective, the 31st TES anticipated issues and problems and developed solutions to overcome them. Most participants agree that the 31st TES's participation was critical to the success of the Global Hawk flight test program and would probably have helped DarkStar as well.

The Global Hawk flight test experience suggests that in future ACTDs, operational users and the test community must be intimately involved in the program along with the ACTD user. The involvement of the test community must be facilitated by the Agreement. Because the involvement of the 31st was not officially recognized in this program, execution problems arose that resulted from a lack of contractual facilitation. The 31st was eventually able to gain the cooperation of the GHSPO, but significant energy had to be expended to get the required changes made.

Resource Constraints and Mission Planning

From its early stages, the D&E test program pushed hard to generate the required number of sorties. This push was likely driven by the shortened length of the overall Phase III. The cumbersome and time-consuming mission planning system required weeks or months from start to finish. The autonomous nature of the system combined with the extremely long flight duration planned for every flight made the process all the more manpower-intensive. To make matters worse, the Air Force Mission Support System (AFMSS), with which the Global Hawk mission support software must interact, was not designed to accommodate long-duration missions with hundreds of way points.

Because of the pace of the D&E program, flight test personnel were involved in five or more mission planning exercises at any given time. The pace of operations in the MUA's crawl phase was challenging given available hardware, system configuration, and personnel resources. Some participants involved in the flight test program felt that the workload required to support the exercises was more than what was considered reasonable to accomplish given the resource limitations.

Both the program's contractor personnel in San Diego and the GHSPO in Dayton were aware of the mismatch between tasking and resources. However, the contractor and government flight test personnel at EAFB adopted a "can-do" attitude and worked extremely hard. The 31st TES was overworked and stated after the fact that they had anticipated a mission planning breakdown. This eventually occurred in the form of the air vehicle 3 taxi accident, which was a direct result of the cumbersome mission planning process combined with the acceleration of the flight test program.

Air Vehicle 3 Postflight Taxi Accident

The postflight taxi accident with air vehicle 3 on December 6, 1999 (41st sortie), resulted from a mission planning failure. The mission was curtailed as a result of a problem with avionics bay temperatures. An early return to base was commanded, putting the system into contingency mode. As a consequence of air traffic congestion at EAFB, Global Hawk was forced to use a secondary contingency plan for landing and taxi. While the landing itself was uneventful, the postflight taxi commands for that particular contingency had not been validated. The mission plan had the air vehicle accelerating and turning right, as it would after takeoff. The air vehicle ended up in the desert with its nose buried in the sand.

The mission planning process failed to detect the problem. As a result of the crash, mission planning validation procedures were changed. The cumbersome mission planning process, along with an increased operating tempo and limited trained personnel, contributed to a heavy burden on Global Hawk test personnel. These were noted as contributing factors in the accident report.[11] The change to AFFTC as RTO was due entirely to the air vehicle 3 mishap and cost the program a three-month slip in an already shortened D&E phase.

Both mishaps during the test program—the destruction of air vehicle 2 and that of air vehicle 3's high-speed taxi—had two significant effects on the ACTD's core objectives. First, each mishap resulted in the loss of the EO/IR onboard sensor; hence, the operational

[11]Accident Investigation Board report, December 6, 1999, excerpted in "Faulty Mission Preparation Cited in December Global Hawk Accident," *Inside the Air Force*, Vol. 11, No. 17, April 28, 2000, pp. 12–13.

demonstration provided no representative EO/IR imagery. Second, the timing of the losses was such that the test program never had more than two flyable aircraft at any given time and was not able to use more than a single aircraft in any demonstration.[12]

Flight Test Responsibility Change

Giving Ryan the lead for flight test activities was not an issue until the air vehicle 3 postflight taxi accident on December 6, 1999. All program participants had accepted the arrangement, and it had proved a workable one until that point. The nature of the accident was such that AFFTC no longer believed it could guarantee safety without being designated as the RTO.

When the RTO dispute arose, the SPO articulated the division of responsibilities that had been in effect throughout the flight test program: Ryan had the lead for flight test program planning and execution and was the home organization of the test director, while the SPO retained accident liability and responsibility for contingency planning and mishap investigation. AFFTC had always held test range safety responsibility. However, neither the Inspector General (IG) nor EAFB would agree to continuing under this structure. The issue was eventually resolved at the three-star level (EAFB/CC and ASC/CC). AFFTC became the RTO on February 7, 2000,[13] and immediately changed the rules on Global Hawk flight testing. Prior to the accident, AFFTC personnel had never expressed safety concerns warranting such changes.

Program participants differed significantly in their opinions on whether the substantial contractor involvement contributed to either of the Global Hawk mishaps. Ryan was not experienced as a flight test lead in an Air Force context. The contractor had a high degree of engineering competence but less experience running an operation in

---

[12]This would be true for most of the D&E exercises even if resources (spare parts, trained personnel) had been increased. It is possible that had there been no resource constraints, air vehicles 1 and 4 and/or air vehicle 5 could have been flown simultaneously toward the end of Phase III.

[13]See memorandum for ASC/RA from HQ AFMC/DO (Brigadier General Wilbert D. Pearson), Subject: Responsible Test Organization Designation, Global Hawk System Test Program, February 7, 2000.

which safety and maintenance are primary concerns. Expertise in these areas lies in the flight test community. Ryan demonstrated a strength in flight test execution but a weakness in flight test planning (which the SPO needed to supplement).

Under the ACTD and OTA constructs, where contractors are given broader responsibilities, it should be required that contractors team with the test community for flight test operations. This did occur in the Global Hawk program; a strong collaborative relationship developed between Ryan and the GHSPO. The GHSPO had between two and ten personnel at each test meeting, taxi, and flight to support testing. The GHSPO had five or more personnel participating at most of the flights. For this type of approach to work, both the contractor and the government must appreciate and accommodate one another's objectives and cultures. Furthermore, the involvement of both the force provider (ACC) and the warfighter (JFCOM) is required early on to shape the system's evolution and concept of operations (CONOPS).

Despite the tension, the loss of several months of flight testing, and the need to change operational procedures, some participants felt that the RTO change did not have a significant impact on the program's ability to accomplish the ACTD objectives. Integration with other organizations (e.g., AFFTC and the FAA) was acknowledged to be somewhat more difficult. Some program participants believe that had AFFTC been the RTO, the air vehicle 3 postflight taxi mishap might have been avoided; because AFFTC is aware of lessons from past programs, it might have gone slower using a more structured process.

The change to AFFTC as RTO had undesirable consequences from JFCOM's perspective: increased bureaucracy, increased rigor required on defining missions, and new and more rules to follow.

Configuration Changes

The system's configuration evolved throughout the D&E phase as the results of continuing nonrecurring engineering activities and lessons from previous flight testing were incorporated into both hardware and software. For the most part, these changes were small and did not affect the conduct of the D&E program.

Block 1 modifications, the first significant hardware configuration changes, were incorporated onto air vehicle 4 before its delivery. These improvements included a second radio altimeter, fuel system improvements, and navigation system improvements (incorporation of the OmniSTAR DGPS navigation system as a replacement for the original LN-211 system used in Global Hawk). In this configuration, air vehicle 4 deployed to Eglin Air Force Base, with a subsequent flight to Portugal as part of the Linked Seas exercise. The flight test team had little experience with the modified air vehicle when these deployments were undertaken, causing some participants to feel that unnecessary risk was taken in deploying the modified configuration without additional testing.

Air vehicle 5 was delivered with the same Block 1 configuration. Air vehicle 3 was updated to the Block 1 configuration during its repair from the runway incident. Air vehicle 1 will also be modified to the Block 1 configuration, bringing all four surviving ACTD residual aircraft to a standard configuration.

The data generated during wing pressure validation sorties (Flights 24 and 25 of air vehicle 1) supported wing redesign on air vehicle 6 to mitigate fuel imbalance. Air vehicle 6, incorporating all the changes mentioned above, is referred to as the Block 2 configuration.

We found no evidence that the configuration changes to either the air vehicle or the ground segment caused significant problems during the ACTD flight test program. Although there were some interoperability and backward compatibility concerns during Phase III, all remaining ACTD air vehicles and ground stations had been brought to the Block 2 configuration by the end of the ACTD program.

Concept of Operations

The Global Hawk system CONOPS has been an area of disagreement between JFCOM and ACC. D&E tests were structured to demonstrate JFCOM's CONOPS, since JFCOM was the ACTD user. The December 1997 ACTD management plan directs JFCOM to validate its CONOPS through field demonstrations, which is precisely what it did. The inherently joint perspective held by JFCOM resulted in the direct dissemination of imagery from the Global Hawk MCE to multiservice exploitation systems. JFCOM found it necessary to keep vehicle control with the Air Force but advocated that real-time sensor control commands—a capability known as dynamic retasking—be allowed to alternate among field organizations as a function of geography, time, and mission/function.

In contrast, the ACC CONOPS, developed late in the ACTD to support post-ACTD planning, called for air vehicle and sensor retasking by Air Force elements only. Imagery users from the other services would have no input regarding what imagery was obtained "on the fly." ACC advocated MCE linkage to Air Force systems only, with imagery selected by the Air Force supplied to the other services secondhand. The initial ACC perspective was Air Force–centric and did not make optimal use of the capabilities already demonstrated by the nascent Global Hawk system.

JFCOM attempted to influence both requirements and CONOPS by demonstrating the feasibility and utility of its CONOPS vision. The Global Hawk system demonstrated multiple dissemination and exploitation links during the D&E exercises, including sensor-to-shooter links and dynamic retasking of the sensor by exploitation users. In other words, the JFCOM CONOPS was demonstrated. ACC was not in a position to demonstrate its CONOPS during the D&E portion of the ACTD.

The CONOPS adopted at the conclusion of the ACTD in large part drives the requirements and future development efforts for the operational system. It will shape the system as a joint asset or as an Air Force asset. ACC is responsible for generating the requirements for the post-ACTD Global Hawk system. The current system was designed to demonstrate—and did demonstrate—a different set of capabilities than those being put forth by ACC.

Military Utility Assessment

The experience of the Predator program[14] was reviewed by participants in the HAE UAV ACTD. The top-down view was that the approach worked—i.e., that Predator provides a useful capability to

---

[14]For a history of the Predator ACTD experience, see Michael R. Thirtle, Robert V. Johnson, and John L. Birkler, *The Predator ACTD: A Case Study for Transition Planning to the Formal Acquisition Process*, MR-899-OSD, Santa Monica: RAND, 1997.

decisionmakers. The bottom-up view was that the system is difficult to operate and sustain. These two views contrast the warfighter as a senior-level decision maker and as the field operator. In the future, both perspectives should provide input to improve the utility and operational suitability of systems.

The MUA process adopted in the Global Hawk program influenced decisions regarding the flow of imagery data and dissemination, requirements generation, and post-ACTD development activities. Different participants believe different things about the value of that information flow. At one extreme, some believed it should dominate future planning regarding requirements and EMD activities. At the other, some advocated almost ignoring that process in future planning because of perceived flaws inherent in the ACTD program structure and in system concept and configuration. This conflict has not been completely resolved. Most program participants did agree that lessons from the ACTD program should be used as a foundation upon which to build future capability.

Experience gained during the D&E exercises allowed for the identification of which performance characteristics really mattered to imagery users and to mission success. The ability to dynamically retask the sensor to take advantage of targets of opportunity turned out to have significant value to the user. This capability was not part of the original design concept but rather emerged as operating experience was gained. Similarly, the number of SAR images taken during a given sortie tended to matter less (within limits) than the quality of those images. Assessing military utility solely on the capabilities derived from the system engineering design (as AFOTEC's initial metrics were) fails to capture other important aspects of military utility that are uncovered as operational experience is gained.

Well before the final D&E sortie, program participants felt that military utility and technical feasibility had been more than adequately demonstrated. Although more flight hours would have been beneficial, it was felt that they were not necessary from a technical perspective. The reason for the high number of flight hours in the original two-year operational demonstration plan was to demonstrate reliability, maintainability, and supportability. While these system attributes were not as well understood as desired by the end of the D&E, the 13-month phase (ten months of operations plus three months of down time due to the air vehicle 3 mishap) was adequate to demonstrate the military utility of the system.

JFCOM's MUA gave Global Hawk high marks in most categories. The final MUA report reflects JFCOM's belief that the system has military utility in its current configuration and potentially greater utility as it evolves and matures.

GLOBAL HAWK SUMMARY

Table 3.1 summarizes the Global Hawk flight test program by phase and air vehicle. Air vehicle 1 was clearly the workhorse of the program, participating in both Phase II and Phase III flight tests. Air vehicles 3–5 participated only in Phase III.

Six outcomes of Global Hawk's ACTD flight test experience are either partially or wholly attributable to its novel acquisition approach:

- The mission planning process was cumbersome and time-consuming. The contractors knew at the time of the Phase II bid that significantly more funds would be required to make the mission planning system suitable for sustained operations. However, because the focus of the ACTD was on demonstrating military utility, which at the time was not well defined and did not specify timely sortie generation, a conscious decision was made to defer this investment. Had mission planning operational suitability been incorporated into a definition of

Table 3.1

Summary of the Global Hawk Flight Test Program by Phase and Air Vehicle (number of sorties/flight hours)

Phase	Air Vehicle 1	Air Vehicle 2	Air Vehicle 3	Air Vehicle 4	Air Vehicle 5	Total
II	12/102.9	9/55.1				21/158
III	13/225.4		9/121.8	11/167.8	4/39	37/554
III+					1/8.5	5/25.1
Total	25/328.3	9/55.1	9/121.8	11/167.8	5/47.5	63/737.1

utility early in the program, more funding might have been committed to it, although perhaps at the expense of other activities.

- The program lacked sufficient resources both for training personnel and for providing adequate spares. This was attributable in part to the reallocation of resources within the program to cover increased nonrecurring engineering activity, and in part to a highly constrained budget throughout the duration of the ACTD.

- The pace of the flight test program was too fast given its cumbersome mission planning process and limited resources. Test personnel were clearly overburdened, which appears to have been a contributing factor in the air vehicle 3 taxi mishap.

- The designation of contractors as the lead for flight test direction, planning, and execution could have resulted in a failed program. Contractors may not have the necessary capabilities, experience, and perspective (culture) to run all aspects of a military test program. The test and operational communities thus took on a large portion of the planning and execution of the flight test program. Their assistance was essential to the accomplishments of the program.

- The differences in perspective between the ACTD and post-ACTD user communities regarding the CONOPS proved to be a serious impediment to the program's transition into the Major Defense Acquisition Program (MDAP) process. The initial CONOPS was generated by the DARPA JPO and was then modified and expanded by JFCOM as part of its responsibility as a designated ACTD user. ACC's CONOPS is similar to current systems in terms of its access to sensor retasking and dissemination pathways; ACC believes that this is what the CINCs want. JFCOM's CONOPS takes advantage of advances in communications and processing technology and adopts a joint orientation. The ACTD D&E phase demonstrated the JFCOM CONOPS; ACC has not demonstrated its CONOPS with respect to Global Hawk.

- Differences between the ACTD and post-ACTD user in operational requirements definition are also inhibiting the program's transition to an MDAP. The extent to which the capabilities of the ACTD configuration—demonstrated through testing—should

> determine the requirements for a post-ACTD system is the underlying issue. The spiral development concept planned for use in post-ACTD development implies that requirements will evolve along with the system's configuration and block upgrades. As a result of this process, early configurations will not have the full capability that ACC, the force provider, desires. Initial drafts of the ORD that is required for all MDAPs were not wholly reflective of the system's demonstrated capabilities and subsequent evolution based on known shortfalls.

Many Global Hawk performance parameters are close to the predicted goals, but some fall short in several significant areas. In particular, a 16 percent increase in empty weight and lower-than-predicted aerodynamic performance resulted in a 20 percent endurance shortfall (32 hours versus 40 hours) and a 7.7 percent shortfall in mission cruise altitude (60 kft versus 65 kft).[15] The ACTD program demonstrated the system's capability for autonomous high-altitude endurance flight. Most communications and data links were demonstrated sufficiently. The SAR sensor can provide high-quality imagery. However, the CGS did not demonstrate control of multiple vehicles; nor was the EO/IR sensor characterized sufficiently.

The time to first flight of Global Hawk was somewhat typical, but the time from first flight to first operational use was extraordinarily short. The system demonstrated operational utility in its current configuration and could be used given contractor support in an operational theater of war.

---

[15]The original DARPA mission profile shows a 3000-nm ingress, a 24-hour on-station segment at 65 kft, and a 3000-nm egress. It is this on-station "cruise" segment that Global Hawk cannot achieve. Global Hawk can achieve an altitude of 65,000 ft for shorter periods of time under certain environmental and weight-related (e.g., fuel remaining) conditions.

Chapter Four

DARKSTAR

Our information on DarkStar is less complete than that for Global Hawk. This section presents available information regarding the DarkStar flight test program.

INITIAL PLAN AND UNDERLYING PHILOSOPHY

Although DarkStar's origins are very different from that of Global Hawk, and although the two HAE UAV programs were originally separate, the DarkStar ACTD flight test program was intended to be similar to that of Global Hawk. Initial plans called for two engineering prototype UAVs to be fabricated for flight test beginning in the fall of 1995. Accompanying the UAVs would be one SAR, one EO sensor, one launch control and recovery station (LCRS), and one interim processing and display system. The last two systems were to be replaced by a CGS in subsequent phases. Phase II was to be a 12-month, contractor-run flight test program.

The December 1997 ACTD management plan states that the objective of Phase II was to partially characterize the system and to improve confidence that the $10 million UFP could be attained.[1] Taking radar cross section measurements was included as an objective during Phase II, reflecting DarkStar's LO design and mission profile.

---

[1]For a complete discussion of the UFP, see Robert S. Leonard and Jeffrey A. Drezner, *Innovative Development: Global Hawk and DarkStar in the HAE UAV ACTD—Program Description and Comparative Analysis,* MR-1474-AF, Santa Monica: RAND, 2001.

The initial (1994) HAE UAV ACTD management plan showed a limited user demonstration for DarkStar lasting approximately 18 months after the Phase II engineering tests. At the time, DarkStar and Global Hawk were on different development schedules, with DarkStar somewhat further ahead in design.

Following the destruction of its first aircraft, DarkStar flight testing was put on hold for more than two years. That delay, combined with Global Hawk flight test program delays, put the systems on roughly consistent schedules for the D&E. The plan for Phase III D&E became a combined 24-month flight test phase using both HAE UAVs. Up to eight additional DarkStar air vehicles and sensor payloads were to be fabricated during Phase III in support of the D&E program, funding permitting.

FLIGHT TEST PROGRAM EXECUTION

DarkStar taxi testing began in January 1996. The first flight of DarkStar air vehicle 1 took place on March 29, 1996, 21 months after the Phase II Agreement was signed. The air vehicle showed significant anomalies, particularly in its "wheelbarrowing" for about 100 yards just prior to liftoff. As discussed previously,[2] LMSW and JPO engineers knew that there were unexplained phenomena and therefore recommended that the second flight be delayed. LMSW and JPO program management backed this decision but were overruled by more senior LMSW and DARPA decisionmakers. DarkStar air vehicle 1 was destroyed just after takeoff on its second flight attempt. Although the primary cause of the crash was related to air vehicle design, several additional factors contributed:

- The accelerated development schedule contributed to insufficient aerodynamic simulation data and modeling, which are usually considered essential when a new aerodynamic configuration is involved.
- Technical risk was significantly underestimated, partially as a result of an overestimation of the value of work on prior LO UAVs.

---

[2]See Drezner, Sommer, and Leonard, *Innovative Management in the DARPA High Altitude Endurance Unmanned Aerial Vehicle Program*, 1999, pp. 76–80.

- The effort was hampered by poor communication and an overall poor relationship between the equal partners in the program: LMSW and Boeing.
- LMSW management pursued an aggressive schedule in an attempt to demonstrate that the Skunk Works remained the world's premier advanced technology development aerospace organization.

The crash of air vehicle 1 resulted in a 26-month flight test hiatus for DarkStar and contributed to increasingly risk-averse behavior in all other segments of the HAE UAV program.

The flight test experience of air vehicle 2 (Figure 4.1) is reproduced here for convenience. The first flight of air vehicle 2, which marked the resumption of Phase II flight testing, occurred on June 29, 1998. Five additional sorties were flown prior to program cancellation in January 1999. While the DarkStar air vehicle 2 flight test program showed the same caution as the initial part of the Global Hawk flight test program, the tone of the after-action reports was somewhat different. DarkStar continued to have major unresolved flight anomalies up to the time of program cancellation. The after-action reports reflected some concern on the part of program engineers; the aerodynamic behavior of the air vehicle continued to show significant differences from what was expected. These differences were not sufficiently explained, although there are some indications that the problems were being brought under control through changes in control software and flight test operations and procedures.

In contrast to Global Hawk, the 31st TES did not participate in the DarkStar flight test program, in compliance with LMSW's wishes. The 31st TES did monitor DarkStar's progress. Just prior to cancellation, LMSW reversed its stance and asked the 31st TES for active support; however, the program was terminated before the 31st TES became involved.

DarkStar's minimal flight hours allowed only limited information to be collected on the flight characteristics of the air vehicle configuration. Although initial sortie rates were the same as those of the Global Hawk program, cumulative flight hours in the first five sorties were less than those of Global Hawk's first five flights. DarkStar completed five sorties in six and one-half months for six total flight

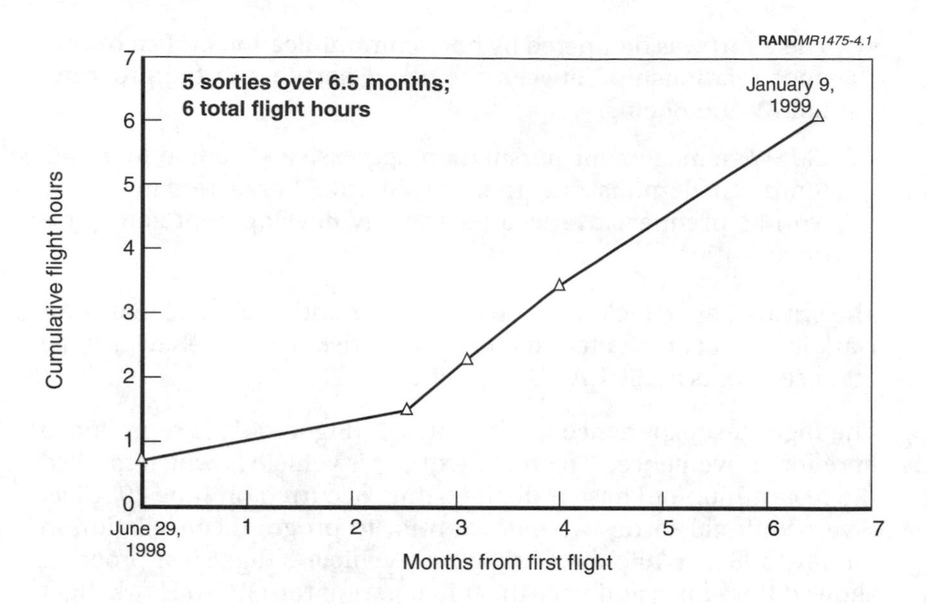

Figure 4.1—DarkStar Air Vehicle 2 Flight Test Program

hours; Global Hawk completed five sorties over seven months, accumulating 20.5 flight hours.

The key question regarding the DarkStar flight test program concerns what program officials learned from it. The program was terminated well before engineering tests had been completed and the flight characteristics of the air vehicle had been confidently understood. DarkStar did not participate in any D&E activities, and no imagery was taken to characterize sensor performance onboard the air vehicle.[3] While the termination decision was explained by future-year funding constraints, flight testing could have continued for the duration of the ACTD at very little cost to the Air Force.

Our reading of DarkStar test results suggests that air vehicle performance would have been considerably less than the stated goals; the aerodynamics of the configuration were clearly not well understood,

---

[3]DarkStar itself never carried the SAR or EO payload in flight.

implying that any early predictions of ultimate mission performance were highly uncertain. DarkStar payloads apparently performed well in tests but were never tested onboard the UAV.

Chapter Five

COMPARISON TO OTHER AIRCRAFT PROGRAMS

This section compares the Global Hawk flight test experience with that of other aircraft programs. The Global Hawk ACTD ended with a system that was more developmentally mature than what is typically seen at the conclusion of an advanced technology demonstrator (ATD) or demonstration/validation (dem/val) program. At the same time, the Global Hawk ACTD did not attain the system maturity typically seen at the conclusion of a traditional EMD program. Because the flight test portion of the Global Hawk ACTD was unique, direct comparison to ATD, dem/val, or EMD flight test programs should be interpreted with caution.

The conditions of flight testing in the Global Hawk program differ considerably from those of other air vehicle types mainly because the system is an autonomous UAV and because it was used in operational demonstrations. Its innovative acquisition approach appears to have had little effect either on the number of sorties or on the accumulation of flight hours for basic and follow-on engineering development and air vehicle checkout flight testing. The innovative acquisition approach does account for earlier user involvement and operational-style testing (the D&E phase) as well as for relatively more substantial participation on the part of the contractors.

An autonomous UAV requires a better understanding of the system earlier in testing because there is no real-time human input or real-time human feedback from the air vehicle as it is tested. By contrast, remotely controlled UAVs have a human in the loop if not onboard and thus gain the advantage of real-time human input. These systems still suffer from a lack of real-time human feedback, as the remote pilot cannot "feel" the air vehicle during test. Manned aircraft have the benefit of both pilot feedback and pilot reactions to identify and mitigate problems during flight testing. The very early operational tests conducted as part of the ACTD are not part of traditional programs, which do not begin operational test until DT has been completed and the system configuration is stable and well understood.

Comparisons with different aircraft types are further complicated by differences in the area and composition of each aircraft type's flight envelope. Fighter, bomber, attack, and cargo are different aircraft types, each generally having a different flight envelope. The Global Hawk mission profile is specifically and narrowly defined, giving the air vehicle a flight envelope with a very small area. Other aircraft types have missions that rely on great aerodynamic flexibility, dictating a much larger-area flight envelope that must be explored during flight test.[1]

Figure 5.1 presents a rough comparison of aircraft flight test programs. Global Hawk's flight test experience is compared with that of several fighter aircraft during their ATD and EMD development phases. We observe a very different pattern of flight-hour accumulation over time in the Global Hawk ACTD. Fighter aircraft tend to accumulate hours much faster as multiple aircraft fly multiple short sorties each month. Global Hawk had only two aircraft in flight-ready status at any one time and generally flew no more than three long-duration sorties per month.

A comparison of the Global Hawk ACTD to prototype programs provides a different perspective.[2] The Attack Aircraft Prototype (AX) program, which involved two pairs of attack prototype aircraft, included a very short, seven-month flight test program. The first five months were conducted by the contractors, and in the next two a

[1]There are no precise comparisons to Global Hawk. Data on the U-2 flight test program, the system in the current inventory that is most similar to Global Hawk, were unavailable.

[2]Data on the Attack Aircraft Prototype (AX) and LWF programs are taken from Giles K. Smith, A. A. Barbour, Thomas L. McNaugher, Michael D. Rich, and William L. Stanley, *The Use of Prototypes in Weapon System Development*, R-2345-AF, Santa Monica: RAND, 1981.

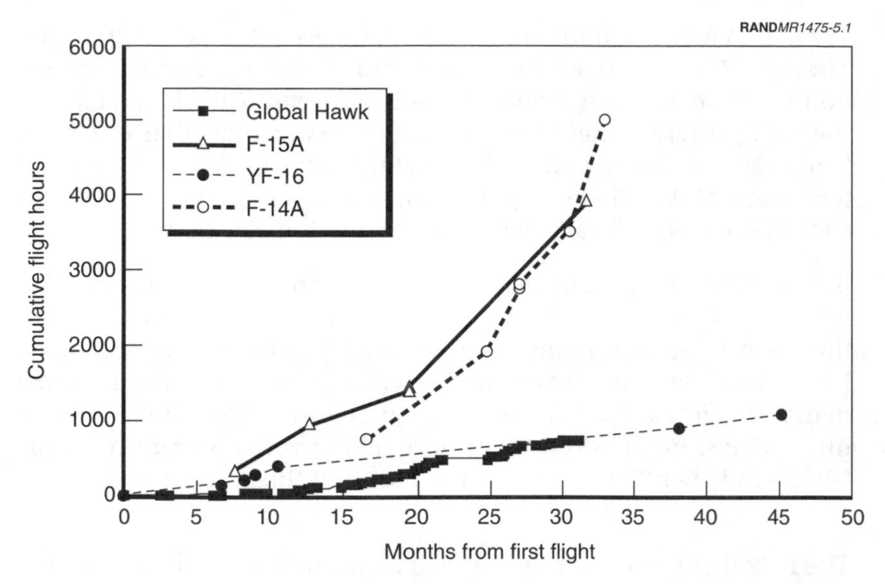

SOURCE: Global Hawk data are from quick-look and flash reports (as of October 2, 2000). Data on the F-14A, F-15A, and YF-16/F-16 were collected by an industry source and validated with other information collected by RAND. The F-14A data include the period from December 1970 through September 1973. The F-15A data include the period from July 1972 through March 1975. The YF-16/F-16 data include both the LWF program and some full-scale development from January 1974 through October 1977.

Figure 5.1—Comparison of Aircraft Flight Test Programs

formal competitive fly-off using Air Force pilots was undertaken. In the first five months, the YA-9 accumulated 162 hours using two aircraft and an additional 145.5 hours (in 123 sorties) in the two-month competitive evaluation. The competing YA-10 accumulated 190 hours in the first five months with its two aircraft and 138.5 hours (87 sorties) in the two-month fly-off. The YA-10s were then used for 16 months during full-scale development (FSD) for follow-on development test and evaluation (DT&E) and initial operational test and evaluation (IOT&E) prior to the low-rate production decision. In contrast, Global Hawk accumulated 20.5 flight hours in five sorties in the first six months of its flight test.

The LWF program of the early 1970s was a highly streamlined effort involving two pairs of prototype fighter aircraft. In 11 months in

1974, the YF-16 accumulated approximately 450 hours in 320 sorties. The YF-17 flew 230 sorties and accumulated approximately 350 hours in a seven-month period that same year. Clearly the pace of the LWF program flight test was much faster than that of Global Hawk. This is due in part to Global Hawk's system type—a large autonomous UAV. Global Hawk requires much more preparation for each mission, specifically in the area of mission planning.

Both competitive prototype flight test programs described above had informal MUAs in the form of participation on the part of operationally oriented test pilots. The use of JFCOM as the architect of the D&E phase of the Global Hawk ACTD, with a focus on formal demonstration of the UAV as a joint warfighting asset and a military utility decision based on that approach, is very different from the "ACC-equivalent" user involvement seen in these comparative programs from the 1970s.

The F-22 flight test program planned an average of 25 flight hours per month during EMD along with the use of multiple aircraft and sorties.[3] Global Hawk can accumulate that in a single sortie. The F-22 also planned to accumulate 1400 hours prior to production authorization but will attain less. The F/A-18E/F logged a total of 4673 flight hours in 3172 flights over 3.5 years of EMD testing at Patuxent River, Maryland, using seven aircraft.[4] Global Hawk has accumulated 737.1 hours in 31 months using five aircraft, never flying more than one at a time.

In general, traditional fighter aircraft EMD flight test programs fly more sorties per month and accumulate more flight hours than did Global Hawk during its ACTD. The main explanation for this difference lies in the vastly larger flight envelope for fighter systems and the need to satisfy myriad requirements, as is traditionally the case in an MDAP-compliant EMD program.

---

[3]See *Aviation Week & Space Technology*, March 27, 2000.

[4]See *Defense News*, June 7, 1999, p. 21.

Figure 5.2 compares the flight test experience of Global Hawk with that of the F-117 EMD program.[5] The two programs show more similarities than do the other comparisons. This pattern is due to flight test constraints, characteristics of the F-117 itself, and the streamlined acquisition approach adopted. Although mission profiles for the F-117 are undoubtedly classified, it is reasonable to assume that the aircraft's flight envelope has a relatively small area. This presumption is driven by the relatively poor aerodynamic characteristics of the F-117, which stem in turn from the aircraft's first-generation LO design.

Like Global Hawk, the F-117 was not held to a long list of technical requirements; instead, the dominant emphasis was on low observability. The program's classified status and subsequent restriction to flight testing only at night slowed the pace at which flight hours

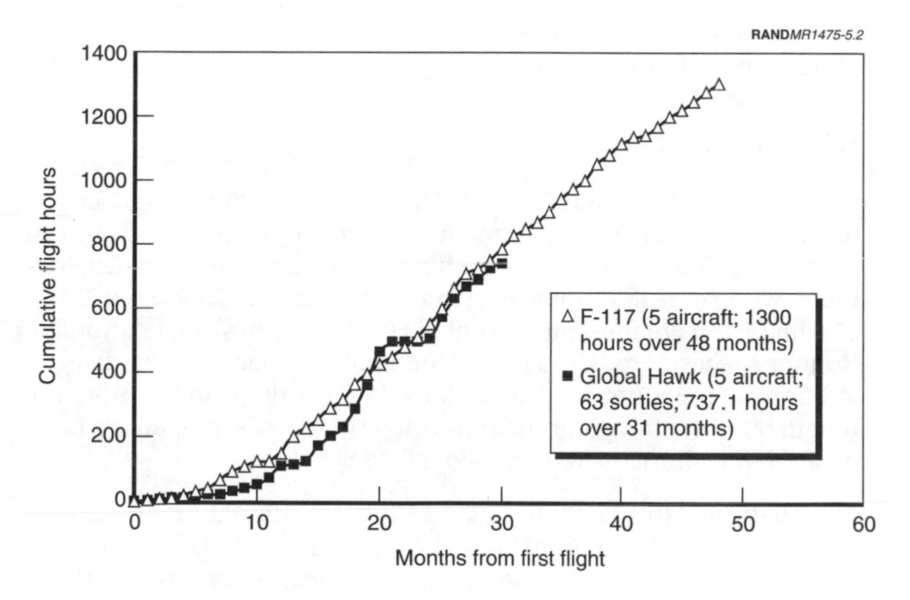

Figure 5.2—Comparison of Global Hawk and F-117 Flight Test Programs

---

[5]See Giles K. Smith, Hyman L. Shulman, and Robert S. Leonard, *Application of F-117 Acquisition Strategy to Other Programs in the New Acquisition Environment,* MR-749-AF, Santa Monica: RAND, 1996.

could be accumulated, much as the mission planning process slowed the Global Hawk program. Similarly, the F-117 generally did not fly all five aircraft in the same month.

The Global Hawk ACTD program has also been described as comparable to the development of the F-117 in culture. The comparison to the F-117 EMD flight test program supports the belief that what was accomplished in the Global Hawk ACTD includes many tasks not usually undertaken until well into an EMD program.

One major difference between the two programs lies in the earlier attention paid to supportability in the Global Hawk program. At the end of the ACTD, 300 of the 900 troubleshooting procedures had been documented. These 300 procedures cover all the usual day-to-day operations of the aircraft. The remaining 600 apply to specialized procedures that are required on a less-than-regular basis. Very little attention was paid to supportability issues during the Have Blue and initial F-117 FSD program.

UAV Comparisons

Boeing's Condor, an autonomous UAV and the indirect predecessor to Global Hawk, flew 141 hours in eight sorties during its flight test program in 1988–1989.[6] Boeing funded most of Condor's development, with some additional funding from DARPA. Condor used two 175-hp piston engines, unlike Global Hawk's use of a single turbofan. Condor's test program did not include payload tests or user demonstrations. Unfortunately, the lack of data for the Condor program, together with the program's composition, does not allow for a meaningful comparison to the Global Hawk.

The Medium-Altitude Endurance (MAE) UAV (Predator) ACTD program was the only other major ACTD program to transition to major program status with operational forces. This program was accelerated even in comparison to the HAE UAV. Its initial flight testing was scheduled for six months in one configuration, with another six-month period planned one year later in a different sensor and com-

---

[6]See Breck W. Henderson, "Boeing Condor Raises UAV Performance Levels," *Aviation Week & Space Technology*, April 23, 1990.

munications configuration. Multiple air vehicles were available to support both test periods. Four D&E exercises were identified in one of the earliest management plans,[7] along with the promise of contractor support in an operational deployment if required.

The development challenge the Predator ACTD faced was much less than that of Global Hawk in several important respects: the Predator is smaller, cheaper, much less complex and capable, and based on an existing system (GNAT-750). However, this comparison illustrates that in the ACTD environment, testing focuses heavily on what the designated user wants in order to assess military utility. This approach implies combined DT/OT without reference to a stable set of system specifications. The Global Hawk experience appears to be similar to that of Predator in this respect.[8]

Similarly, the Tactical UAV (TUAV) ACTD program identified the operational test organizations in the Army, Marine Corps, and Navy as the lead test agencies, just as AFOTEC was the lead test agency in the HAE UAV. This reflects the operational demonstration focus of ACTDs, as opposed to the demonstration of mature technical development.

A Different Metric: Early Identification of Major Design Flaws

Another metric for comparison concerns the ability of the test program to find significant technical or performance flaws prior to the commitment of major funding (e.g., production). Earlier RAND work on the adequacy of test programs examined past aircraft programs to better understand the point at which significant design flaws affecting mission capability are discovered during flight test.[9] Table 5.1 provides data on a sample of programs. The overall conclusions suggest that such problems (e.g., the F-117 tail, C-5A wing fatigue, and

---

[7]Defense Advanced Research Projects Agency, *HAE UAV ACTD Management Plan*, December 15, 1994.

[8]See Thirtle, Johnson, and Birkler, *The Predator ACTD: A Case Study for Transition Planning to the Formal Acquisition Process*, 1997.

[9]See Giles K. Smith, "Use of Flight Test Results in Support of F-22 Production Decision," internal document, Santa Monica: RAND, 1994; and Giles K. Smith, "The Use of Flight Test Results in Support of High-Rate Production Go-Ahead for the B-2 Bomber," internal document, Santa Monica: RAND, 1991.

Table 5.1

Major Problems Revealed During Flight Test

Program	Major Problem Identified	Percentage of Testing Complete
C-5A	Static test failure in wing root	20
	Hydraulic leaks/engine	25
	Landing gear mechanisms	2
	Multimode radar deficiencies	15
	Wing fatigue problem	40–50
B-1A	Weapon bay acoustics	5
	Shock-induced oscillations	60
	Horizontal stabilizer fatigue	10
B-1B	Defensive avionics	10
	Terrain-following radar	10
F-117	Tail size	1
	Wing structure	10
	Infrared Attack and Designation System	10
	Rudder	50
F/A-18A	Excessive drag	15
	Bulkhead fatigue cracks	10
	Inadequate roll rate	20

SOURCE: Giles K. Smith, "Use of Flight Test Results in Support of F-22 Production Decision," internal document, Santa Monica: RAND, 1994.

B-1B defensive avionics) are rare and usually occur early in the test program. Global Hawk was clearly well past that point by the end of the ACTD; DarkStar illustrates the point.

Chapter Six

TEST PROGRAM COMPOSITION AND THE TRANSITION TO MDAP

Some program participants believe that neither the content of the flight test program (what was done) nor its approach (how it was done) was greatly affected by its acquisition strategy. These participants assert that test program management would have been similar under a traditional acquisition approach. Indeed, evidence suggests that the dominant influence on the pace and structure of the test program was the nature of the system; until the HAE UAV ACTD program, very little experience had been accumulated with large autonomous UAVs. System characteristics determined both the profile in which flight hours were accumulated over time and the types of tests that were conducted (e.g., minimal envelope expansion testing). However, the level of system maturity attained at the conclusion of the ACTD was akin to being partway through an EMD flight test program. This creates uncertainty in structuring the post-ACTD flight test program.

The acquisition approach clearly influenced some key elements of the test program: the increased contractor involvement and the early operational testing in the form of user demonstrations.

For Global Hawk, operational experience was gained at a fraction of the resources and flight hours initially presumed to be required; available assets and cumulative flight hours were sufficient to demonstrate military utility. The inherent flexibility of the system was poorly understood until the D&E phase. The CONOPS evolved as flight experience was gained (a primary purpose of an ACTD). The relative importance of imaging rate versus sensor and air vehicle retasking evolved; retasking turned out to be more important to the user community.

A less obvious result of the acquisition approach stemmed from the program's lack of resources. Cost increases in early development stages, long before flight testing began, led to a reduction in assets during the ACTD. The ACTD process put the total effort on a strict overall schedule, effectively placing it on a fixed budget as well. The result was not only a shortened D&E but also fewer assets available for that D&E. Had early development efforts gone more as planned, more assets would have participated in a longer D&E phase. This would almost certainly have allowed more flight hours to be accumulated, thus establishing the criteria desired by JFCOM. Simultaneous operation of multiple air vehicles would almost certainly have been demonstrated as well. Finally, more assets would have included more sensor suites, and thus the EO/IR sensor that was never characterized would almost certainly have been characterized. Through these results of a notional extended D&E phase, some of ACC's current issues could have been addressed and possibly resolved, thereby facilitating the transition to an MDAP.

Even slightly more resources would have made a significant difference. A traditional approach usually includes more resources in the areas of spare parts and trained personnel. The relatively low budget ACTD program led to a parts shortage; other aircraft were commonly cannibalized for parts. Increased spares and other subsystem assets might have increased flight hours. With the severely limited resources at hand, flight operations could not be sustained on more than one air vehicle at a time owing to a lack of trained maintenance and operations personnel. Shortages of both parts and trained personnel meant that air vehicle 5 could not be flown at EAFB while air vehicle 4 was deployed to Eglin.

The Global Hawk D&E program was not expected to accomplish the full set of operational test and evaluation (OT&E) tests required for an MDAP. However, engineering tests performed during the ACTD should satisfy some DT requirements. OT&E should certainly benefit from the operational experience gained during Global Hawk's D&E flight test program.

Following the pattern established during the ACTD, program documents indicated that a combined DT/OT program will be implemented in follow-on phases. A test and evaluation management plan (TEMP) will be developed and approved. AFOTEC will perform an operational assessment (OA) by leveraging Phase III experience. IOT&E/follow-on operational test and evaluation (FOT&E) to evaluate ORD compliance will be initiated when development warrants.[1]

The EMD phase will not start with a new air vehicle that corrects identified "deficiencies." Instead, the GHSPO intends to use the ACTD configuration to support the EMD program until the next air vehicles (Block 5) become available. All program participants advocate a continued operational flavor to EMD flight tests. Some have recommended roughly three flights per quarter in one D&E exercise to remain visible in the operational world. Funding for operational demonstration flights during EMD had not been assured.

The configuration evolved throughout the program, as is the norm in the early development stages of a traditional approach. What was different in the HAE UAV ACTD program was that there was no stable system specification to test against and to provide input into requirements generation. In a traditional approach, a firm requirement is translated into a system specification prior to entering EMD. In contrast, one purpose of the HAE UAV ACTD was to improve our understanding of what the requirements and CONOPS should be for an autonomous UAV in an ISR mission role. The ACTD test program was more about understanding the capabilities of the system that was designed than about demonstrating a given level of performance corresponding to a system specification. In a traditional program, these priorities are reversed.

This basic difference between an ACTD test program and a traditional approach is one of the key drivers of the challenges facing program officials as Global Hawk transitions to an MDAP using a more traditional approach. The operational experience gained during the Global Hawk test program represents useful information regarding CONOPS and requirements for a system of this type. The development test experience helped characterize the capabilities of the cur-

---

[1]C4ISP briefing, June 7, 2000.

rent system and identify areas in which improvements are needed. If used judiciously, these two sets of information can vastly improve the efficacy of the post-ACTD (EMD) test program.

Appendix A

GLOBAL HAWK FLIGHT TEST DATA

Table A.1 contains information on the Global Hawk flight test program. Data include sortie numbers, sortie dates, flight hours, objectives, results of the mission, and additional comments regarding significant events during the test. These data were used to create the charts used in Chapters Two and Four.

The data in the table were derived from the flash reports and quick-look reports that documented the results of each sortie.

Table A.1

Global Hawk Flight Test Data[a]

Sortie Number	Date	Air Vehicle	Official Sortie Designation	Flight Time (hours)	Cumulative Flight Hours	Phase	Objective	Results	Comments
1	2/28/98	AV01	AW-01	0.95	0.95	II	Functional checkout: airworthiness first flight	Most primary objectives met	Landing gear would not lock up; ECS low temperature resulted in shortening of mission by 11 minutes.
2	5/10/98	AV01	AW-02	2.4	3.4	II	Functional checkout: demonstrate basic flight controls and systems; instrument calibration	Objectives met	LRE-MCE handoff demonstration; 150 knots current air speed (KCAS), 41,000-ft mean sea level (MSL).
3	5/30/98	AV01	AW-03	5.35	8.7	II	Functional checkout: expand flight envelope and demonstrate systems/functions	Objectives met	Brake problem resulted in stopping slightly off the runway; TO 0701 LD 1222.

Table A.1—continued

Sortie Number	Date	Air Vehicle	Official Sortie Designation	Flight Time (hours)	Cumulative Flight Hours	Phase	Objective	Results	Comments
4	8/29/98	AV01	AW-04	5.7	14.4	II	Functional checkout: expand flight envelope; demonstrate FL600	Most primary objectives met	Minor discrepancies in some systems; FL593 reached (61,000 inertial); TO 0835 LD 1417.
5	9/17/98	AV01	AW-05	6.1	20.5	II	Functional checkout: expand flight envelope; demonstrate FL600 full fuel load	Most primary objectives met	Flown at night, no chase aircraft; IMMC A failed; UHF SATCOM failed again; first interaction with FAA ATC.
6	10/29/98	AV01	AW-06	9.7	30.2	II	Medium endurance		Executed missed approach.
7	11/20/98	AV02	AV02-01	3.1	33.3	II	Functional checkout: first flight; evaluate core airworthiness	Objectives partially met	Shortened by 0.5 hour owing to sticking outboard spoiler; link with MCE could not be established.

Table A.1—continued

Sortie Number	Date	Air Vehicle	Official Sortie Designation	Flight Time (hours)	Cumulative Flight Hours	Phase	Objective	Results	Comments
8	12/4/98	AV02	AV02-02	3.3	36.6	II	Functional checkout: refly of first flight	Objectives mostly met	Wideband communications demonstrated; aircraft taxied off runway autonomously after landing; CDL failure detected prior to takeoff.
9	12/19/98	AV02	AV02-03	7.4	44.0	II	Functional checkout: general and payload objectives	Objectives partially met	General objectives mostly met; no ISS payloads met; reached FL583; ISS shut down; unsuccessful reboot attempts—temperature fell below lower limit.

Table A.1—continued

Sortie Number	Date	Air Vehicle	Official Sortie Designation	Flight Time (hours)	Cumulative Flight Hours	Phase	Objective	Results	Comments
10	1/22/99	AV02	AV02-04	6.4	50.4	II	Payload test	Objectives partially met	General objectives mostly met; some ISS objectives met; imagery taken for the first time (poor quality).
11	1/28/99	AV01	AV01-07	1.7	52.1	II	Functional checkout: lightweight; expand sortie duration envelope	Objectives not met	Engine anomalies due to software problem indicated early RTB: was to evaluate fix to wing fuel transfer discovered in August.
12	2/2/99	AV01	AV01-08	10.1	62.2	II	Functional checkout: refly AV1-07 mission/endurance	Objectives mostly met	Reached 64,700 ft MSL; hydraulic system failure cut flight short; emergency backup systems worked.

Table A.1—continued

Sortie Number	Date	Air Vehicle	Official Sortie Designation	Flight Time (hours)	Cumulative Flight Hours	Phase	Objective	Results	Comments
13	2/20/99	AV02	AV02-05	12.5	74.7	II	ISS basic functionality and characteristics	Objectives met	Reached 61,000 ft; 144 of 156 planned sensor events; continued improvement in Ku-band communications.
14	2/28/99	AV01	AV01-09	18.1	92.8	II	Demonstrate endurance, towed decoy, fuel balance below 5000 lb fuel; basically refly of 2/2/99 mission	Objectives met	Except decoy objective; reached 66,400 ft MSL.
15	3/11/99	AV02	AV02-06	0.7	93.5	II	ISS basic functionality and characteristics	Objectives not met	Aborted: landing gear fault during autonomous retraction.
16	3/13/99	AV02	AV02-07	15.2	108.7	II	Payload functionality test for incentive scoring; continued ISS checkout	Objectives met	Simultaneous SAR and EO spot imaging.

Table A.1—continued

Sortie Number	Date	Air Vehicle	Official Sortie Designation	Flight Time (hours)	Cumulative Flight Hours	Phase	Objective	Results	Comments
17	3/26/99	AV02	AV02-08	5.8	114.5	II	Payload functionality checkout	Objectives not met	Mission terminated early owing to poor SAR performance.
18	3/29/99	AV02	AV02-09	0.7	115.2	II	Payload functionality test	Objectives not met	Air vehicle crash owing to initiation of auto destruct by transmission from Nellis Air Force Base.
19	5/18/99	AV01	AV01-10	7.8	123.0	II	Prepare for payload integration/ wideband link evaluation	Objectives met	First flight after crash.
20	6/3/99	AV01	AV01-11	27.2	150.2	II	SAR sensor checkout/long endurance		

Table A.1—continued

Sortie Number	Date	Air Vehicle	Official Sortie Designation	Flight Time (hours)	Cumulative Flight Hours	Phase	Objective	Results	Comments
21	6/11/99	AV01	AV01-12	7.8	158.0	II	Exercise SAR	Objectives met	Second time AV01 carried SAR.
22	6/19/99	AV01	AV01-13	12.8	170.8	III	D&E—Roving Sands 1	Objectives met	Operated in NAS under FAA control for majority of mission; use of encrypted links for C^2 and imagery; demonstrated quick turnaround time; first in crawl series.
23	6/26/99	AV01	AV01-14	2.0	172.8	III	D&E—Roving Sands 2	Objectives not met	Curtailed: SAR would not boot up.
24	6/27/99	AV01	AV01-15	17.2	190.0	III	D&E—Roving Sands 2B	Objectives met	Operated in NAS under FAA control for majority of mission; use of encrypted links for C^2 and imagery; demonstrated quick turnaround time.

Table A.1—continued

Sortie Number	Date	Air Vehicle	Official Sortie Designation	Flight Time (hours)	Cumulative Flight Hours	Phase	Objective	Results	Comments
25	7/15/99	AV01	AV01-16	12.9	202.9	III	D&E—Extended Range 1-1	Objectives met	Third D&E Phase III flight; 91 of 119 planned sensor events accomplished; imagery disseminated.
26	7/27/99	AV01	AV01-17	23.9	226.8	III	D&E—Extended Range 1-2	Objectives met	Continued expansion and demonstration of Global Hawk capabilities: different imagery, tasking, links, etc.
27	8/12/99	AV03	AV03-01	4.0	230.8	III	Functional checkout: first flight, evaluate core airworthiness	Objectives met	

Table A.1—continued

Sortie Number	Date	Air Vehicle	Official Sortie Designation	Flight Time (hours)	Cumulative Flight Hours	Phase	Objective	Results	Comments
28	8/30/99	AV01	AV01-18	25	255.8	III	D&E—Extended Range 2/JEFX/CAX	Objectives met	66,000 MSL; 5 ISS restarts during D&E phase; 3 during engineering phase of flight.
29	9/9/99	AV01	AV01-19	18.8	274.6	III	D&E—CAX 99-10 U.S. Marine Corps Exercise	Objectives met	Sixth D&E participation flight.
30	9/16/99	AV03	AV03-02	12.0	286.6	III	Functional checkout: continue airworthiness	Objectives met	LRE CCO provided by 31st TES.
31	10/4/99	AV01	AV01-20	25.3	311.9	III	D&E—Extended Range U.S. Navy Seals	Objectives met	Except GMTI; Extended Range 3-01 exercise; first in walk series.

Table A.1—continued

Sortie Number	Date	Air Vehicle	Official Sortie Designation	Flight Time (hours)	Cumulative Flight Hours	Phase	Objective	Results	Comments
32	10/8/99	AV01	AV01-21	24.8	336.7	III	D&E—Extended Range 3-02 U.S. Navy Seals and CAS	Objectives met	
33	10/19/99	AV01	AV01-22	24.8	361.5	III	D&E—Extended Range 4-01 Alaska	Objectives met	Except Delta V; Extended Range 4; tenth dedicated D&E exercise.
34	10/25/99	AV01	AV01-23	25.5	387.0	III	D&E—Extended Range 4-02 Alaska	Objectives met	Extended Range 4, second flight; operation outside CONUS; 23.8 hours ISS continuous operations.
35	10/30/99	AV03	AV03-03	3.0	390.0	III	Functional checkout: ISS	Objectives not met	Early RTB due to AC generator anomaly.
36	11/4/99	AV03	AV03-04	24.6	414.6	III	Functional checkout: ISS	Objectives met	AV03 cleared for participation in D&E.

Table A.1—continued

Sortie Number	Date	Air Vehicle	Official Sortie Designation	Flight Time (hours)	Cumulative Flight Hours	Phase	Objective	Results	Comments
37	11/9/99	AV03	AV03-05	17.4	432.0	III	D&E—Desert Lightning II	Objectives met	Notes 434 total test hours in Global Hawk program to date; 230.5 hours flown since June 1999.
38	11/13/99	AV03	AV03-06	9.2	441.2	III	D&E—Desert Lightning II	Objectives not met	Curtailed following IMMC A failure due to low temperature in avionics compartment.
39	11/17/99	AV03	AV03-07	19.8	461.0	III	D&E—Desert Lightning II	Objectives met	14 D&E sorties flown since June 1999.
40	12/3/99	AV03	AV03-08	22.0	483.0	III	D&E—JTF-6 Sortie 1	Objectives met	
41	12/6/99	AV03	AV03-09	9.8	492.8	III	D&E—JTF-6 Sortie 2	Objectives not met	Early RTB; postflight taxi "event."

Table A.1—continued

Sortie Number	Date	Air Vehicle	Official Sortie Designation	Flight Time (hours)	Cumulative Flight Hours	Phase	Objective	Results	Comments
42	3/11/00	AV01	AV01-24	6.6	499.4	III	Engineering test; demonstrate LRE 2, MRE 2; wing pressure validation	Objectives met	62.3 kft; no sensor suite; first sortie with AFFTC as RTO.
43	3/17/00	AV01	AV01-25	5.8	505.2	III	Engineering test: wing pressure validation (2)	Objectives met	
44	3/24/00	AV04	AV04-01	4.1	509.3	III	Functional checkout: first flight AV04	Objectives met	
45	3/30/00	AV04	AV04-02	4.5	513.8	III	Functional checkout	Objectives not met	Early RTB due to INS anomaly; plan was 11.2 hours.
46	4/4/00	AV04	AV04-03	5.4	519.2	III	Functional checkout	Objectives not met	Early RTB due to low fuel temperature; plan was 24 hours.
47	4/11/00	AV04	AV04-04	24.2	543.4	III	Functional checkout	Objectives met	Functional validation of LRE 3 and MCE 2 during flight.

Table A.1—continued

Sortie Number	Date	Air Vehicle	Official Sortie Designation	Flight Time (hours)	Cumulative Flight Hours	Phase	Objective	Results	Comments
48	4/14/00	AV04	AV04-05	31.5	574.9	III	Functional checkout	Objectives met	Maximum altitude 65,100 ft; testing Ku-band satellite links for East Coast deployment.
49	4/20/00	AV04	AV04-06	10.5	585.4	III	D&E deployment to Eglin Air Force Base	Objectives met	Resumption of D&E: LREs and MRE located on East Coast; FAA coordination.
50	5/8/00	AV04	AV04-07	28.0	613.4	III	D&E—Linked Seas - 1	Objectives met	First trans-ocean flight; first East Coast launch and recovery; first NATO demonstration; first Global Hawk operation in international airspace.
51	5/11/00	AV04	AV04-08	14.1	627.5	III	D&E—Linked Seas - 2	Objectives partially met	UHF satellite link problems.

Table A.1—continued

Sortie Number	Date	Air Vehicle	Official Sortie Designation	Flight Time (hours)	Cumulative Flight Hours	Phase	Objective	Results	Comments
52	5/18/00	AV04	AV04-09	22.5	650.0	III	D&E—JTFEX00-1	Objectives met	
53	5/19/00	AV04	AV04-10	14.7	664.7	III	D&E—JTFEX00-2	Objectives partially met	Early RTB—IMMC failure.
54	6/19/00	AV04	AV04-11	8.3	673.0	III	Redeployment to EAFB	Objectives partially met	ISS-IMMC interface problems.
55	6/30/00	AV05	AV05-01	0.8	673.8	III	Functional checkout	Objectives partially met	Early RTB—failure in EO/IR LN-100 navigator.
56	7/7/00	AV05	AV05-02	3.0	676.8	III	Functional checkout	Objectives mostly met	Refly of previous mission; flight curtailed by 1 hour due to failure of ruddervator actuator.
57	7/12/00	AV05	AV05-03	11.4	688.2	III	Functional checkout	Objectives met	SAR checkout.

Table A.1—continued

Sortie Number	Date	Air Vehicle	Official Sortie Designation	Flight Time (hours)	Cumulative Flight Hours	Phase	Objective	Results	Comments
58	7/19/00	AV05	AV05-04	23.8	712.0	III	Functional checkout	Objectives met	Final Phase III flight.
59	8/4/00	AV05	AV05-05	8.5	720.5	III+	Payload configuration functional checkout	Objectives partially met	First test flight in support of Australian deployment; ISS would not boot in new mode.
60	8/11/00	AV05	AV05-06	3.8	724.3	III+	SAR maritime mode evaluation	Objectives met	Preparation for Australian deployment.
61	8/31/00	AV05	AV05-07	5.8	730.1	III+	Refly 5-05 VPH GMTI	Objectives mostly met	Test for video phase history (VPH), ISS reconfigure from VPH to legacy mode.

Table A.1—continued

Sortie Number	Date	Air Vehicle	Official Sortie Designation	Flight Time (hours)	Cumulative Flight Hours	Phase	Objective	Results	Comments
62	9/12/00	AV05	AV05-08	7.0	737.1	III+	Support JEFX time-critical targeting	Objectives met	JEFX first sortie.
63	9/14/00	AV05	AV05-09	0	737.1	III+	Support JEFX time-critical targeting	Objectives not met	Mission abort prior to takeoff—exceed cross-track error criteria on rollout.

SOURCE: HAE UAV JPO flash reports, quick-look reports, and flight test summary data charts.

[a]FL = flight level; ECS = environmental control system; TO = takeoff; LD = landing; ATC = air traffic control; CDL = command data link; ISS = integrated sensor suite; RTB = return to base; NAS = national air space; GMTI = ground moving target indicator; INS = inertial navigation system.

Appendix B

DARKSTAR FLIGHT TEST DATA

Table B.1 contains information on the DarkStar flight test program. Data include sortie numbers, sortie dates, flight hours, objectives, results of the mission, and additional comments regarding significant events during the test. These data were used to create the charts used in Chapter Three.

The data in the table were derived from the flash reports and quick-look reports that documented the results of each sortie.

Table B.1

DarkStar Flight Test Data

Sortie Number	Date	Air Vehicle	Official Sortie Designation	Flight Time (hours)	Cumulative Flight Hours	Phase	Objective	Results	Comments
1	3/29/96	AV01				II	Airworthiness, instrument calibration		“Wheelbarrowing” down runway; discrepancy between model predictions and flight test data.
2	4/22/96	AV01				II	Airworthiness, instrument calibration	Objectives not met	CRASH.
3	6/29/98	AV02	696 F2-1	0.73	0.73	II	Airworthiness, instrument calibration	Objectives met	Oscillation and climb rate problems.

Table B.1—continued

Sortie Number	Date	Air Vehicle	Official Sortie Designation	Flight Time (hours)	Cumulative Flight Hours	Phase	Objective	Results	Comments
4	9/13/98	AV02	696 F2-2	0.77	1.50	II	Pitch oscillation investigation	Objectives met	Wing loads near limit during flight and landing; brake pulsing; climb rate problem; oscillation better.
5	9/30/98	AV02	696 F2-3	0.77	2.26	II	Mission replan demonstration; fleet SATCOM demonstration; flight envelope expansion	Objectives not met	Oscillation continued; mission abort decision after fault message received on replan execute command.
6	10/27/98	AV02	696 F2-4	1.17	3.43	II	Fleet SATCOM demonstration; envelope expansion	Objectives partially met	Initial climb rate still problem.

Table B.1—continued

Sortie Number	Date	Air Vehicle	Official Sortie Designation	Flight Time (hours)	Cumulative Flight Hours	Phase	Objective	Results	Comments
7	1/9/99	AV02	696 F2-5	2.63	6.06	II	Envelope expansion	Objectives not met	Brake system continued to fail; climb rate; aircraft kept reaching flight limits on attempted envelope expansion tests; flight termination signal tested on ground due to detection of spurious signal from elsewhere on range.
8	1/29/99	AV02	696 F2-6	n/a	n/a	II	n/a	n/a	Stop work order/program termination.

SOURCE: HAE UAV JPO flash reports, quick-look reports, and flight test summary data charts.

BIBLIOGRAPHY

"ACTD Transition Guidelines: Executive Summary," available at www.acq.osd.mil/actd/.

"Air Force Lab Pushes UAVs for AWACS, JSTARS, RIVET Joint Missions," *Inside the Air Force,* July 21, 2000, pp. 15–17.

"Air Force to Appeal Senate Cut to Global Hawk Advance Procurement," *Inside the Air Force,* June 2, 2000, pp. 3–4.

"Air Staff to Brief Peters on Accelerating Global Hawk, Retiring U-2," *Inside the Air Force,* July 21, 2000, pp. 11–12.

"Australia Eyes UAV," *Aviation Week & Space Technology,* March 1, 1999, p. 35.

"End of UAV Demos Leaves ASC Wondering How to Budget for the Vehicles," *Inside the Air Force,* December 11, 1998, p. 7.

"Faulty Mission Preparation Cited in December Global Hawk Accident," *Inside the Air Force,* Vol. 11, No. 17, April 28, 2000, pp. 12–13.

"Global Hawk Incident Prompted Changes: Military Alters Communications Management Rules After UAV Crash," *Inside the Air Force,* January 28, 2000, p. 3.

"Global Hawk 2 Flight Sets Stage for Airborne Sensor Tests," *Aviation Week & Space Technology,* November 30, 1998, p. 32.

"Global Hawk UAV Crash Linked to Test Error," *Defense News,* January 10, 2000.

"Improper Mission Computer Input Said to Be Key Factor in UAV Crash," *Inside the Air Force*, March 17, 2000, p. 14.

"Introduction to ACTDs," available at www.acq.osd.mil/actd.

"Let's Make a Deal," *Aviation Week & Space Technology*, August 30, 1999, p. 21.

"Out, DarkStar," *Aviation Week & Space Technology*, February 1, 1999, p. 27.

"Pentagon Argues Global Hawk Cut Will Delay Fielding by One Year," *Inside the Air Force*, July 28, 2000, pp. 15–17.

"Poor Communications Management Cited in Global Hawk UAV Crash," *Inside the Air Force*, January 7, 2000, pp. 9–10.

"RQ-4A Global Hawk Unmanned Aerial Vehicle Accident," *Inside the Air Force*, April 28, 2000, pp. 13–15.

"Senate Authorizers Want to Slow Global Hawk UAV Procurement Plans," *Inside the Air Force*, May 19, 2000, pp. 7–8, 13, and 23–27.

"Senators Push for Global Hawk to Explore Airborne Surveillance Role," *Inside the Air Force*, May 12, 2000, pp. 5–6.

"Tenter, De Leon to Determine Future Airborne Recon Mix, Funding Needs," *Inside the Air Force*, August 4, 2000, p. 3.

Advanced Systems and Concepts, *ACTD Master Plan*, CD-ROM, September 2000.

Asker, James R., "Let's Make a Deal," *Washington Outlook*, August 30, 1999, p. 24.

Atkinson, David, "Global Hawk Crash Delays Demos," *Defense Daily*, April 1999.

Barraco Klement, Mary Ann, "Agile Support Project—Global Hawk Program: Rapid Supply, Responsive Logistics Support for Next-Generation UAVs," *PM*, January–February 1999, pp. 66–70.

Birkler, John, Giles Smith, Glenn A. Kent and Robert V. Johnson, *An Acquisition Strategy, Process, and Organization for Innovative Systems*, Santa Monica: RAND, MR-1098-OSD, 2000.

Congressional Budget Office, "The Department of Defense's Advanced Concept Technology Demonstrations," Washington, D. C., September 1998.

de France, Linda, "UAVs Hold Key to Future Conflicts, Kosovo Air Commander Says," *Aerospace Daily,* November 15, 2000.

Dornheim, Michael A., "Destruct System Eyed in Global Hawk Crash," *Aviation Week & Space Technology,* April 5, 1999, p. 61.

Drezner, Jeffrey A., Geoffrey Sommer, and Robert S. Leonard, *Innovative Management in the DARPA High Altitude Endurance Unmanned Aerial Vehicle Program: Phase II Experience,* MR-1054-DARPA, Santa Monica: RAND, 1999.

Eash, Joseph, "ACTDs Link Speed, Technology for U.S. Forces," *Defense News,* September 20, 1999.

Eash, Joseph, "Two Major UAVs Show Defense Program's Success," *Aviation Week & Space Technology,* August 7, 2000, p. 74.

Fulghum, David A., "Long-Term Stealth Project Gets the Ax," *Aviation Week & Space Technology,* May 24, 1999, pp. 77–78.

Fulghum, David A., "Low-Cost Mini-Radar Developed for UAVs," *Aviation Week & Space Technology,* October 9, 2000, p. 101.

Fulghum, David A., "Recce Funding Increase Pits U-2 Against Global Hawk," *Aviation Week & Space Technology,* September 27, 1999, p. 37.

Fulghum, David A., "Will New Elusive Craft Rise from DarkStar?" *Aviation Week & Space Technology,* February 22, 1999, pp. 27–28.

Fulghum, David A., and Robert Wall, "Global Hawk Gains Military Endorsement," *Aviation Week & Space Technology,* September 18, 2000, p. 34.

Fulghum, David A., and Robert Wall, "Global Hawk Gains Military Endorsement," *Aviation Week & Space Technology,* September 18, 2000, pp. 34–36.

Fulghum, David A., and Robert Wall, "Global Hawk Snares Big Break," *Aviation Week & Space Technology,* October 23, 2000, p. 55.

Fulghum, David A., and Robert Wall, "Long-Hidden Research Spawns Black UCAV," *Aviation Week & Space Technology*, September 25, 2000, pp. 28–30.

Fulghum, David A., and Robert Wall, "New Missions, Designs Eyed for Global Hawk," *Aviation Week & Space Technology*, November 20, 2000, p. 57.

Harmon, Bruce R., Lisa M., Ward, and Paul R. Palmer, *Assessing Acquisition Schedules for Tactical Aircraft*, Alexandria, VA: Institute for Defense Analyses, IDA Paper P-2105, 1989.

Henderson, Breck W., "Boeing Condor Raises UAV Performance Levels," *Aviation Week & Space Technology*, April 23, 1990.

Hundley, Richard O., "DARPA: Technology Transitions: Problems and Opportunities," internal document, Santa Monica: RAND, June 1999.

Johnson, Robert V., and John Birkler, *Three Programs and Ten Criteria: Evaluating and Improving Acquisition Program Management and Oversight Processes Within the Department of Defense*, MR-758-OSD, Santa Monica: RAND, 1996.

Johnson, Robert V., and Michael R. Thirtle, "Management of Class III Advanced Concept Technology Demonstration (ACTD) Programs: An Early and Preliminary View," internal document, Santa Monica: RAND, July 1997.

Leonard, Robert S., and Jeffrey A. Drezner, *Innovative Development: Global Hawk and DarkStar in the HAE UAV ACTD—Program Description and Comparative Analysis*, MR-1474-AF, Santa Monica: RAND, 2001.

Leonard, Robert S., Jeffrey A. Drezner, and Geoffrey Sommer, *The Arsenal Ship Acquisition Process Experience: Contrasting and Common Impressions from the Contractor Teams and Joint Program Office*, MR-1030-DARPA, Santa Monica: RAND, 1999.

Mann, Paul, "Joint Ops Make Gains, but 'Jointness' Lags," *Aviation Week & Space Technology*, April 10, 2000, p. 27.

McMichael, William H., "Global Hawk Teams with U.S. Carrier Group," *Defense News*, June 19, 2000, p. 24.

Moteff, John D., "IB1022: Defense Research: DOD's Research, Development, Test, and Evaluation Program," *CRS Issue Brief for Congress*, August 13, 1999.

Mulholland, David, "Pentagon Cancels DarkStar UAV, Pursues Global Hawk," *Defense News*, 1999.

Pae, Peter, "Military Is Sold on Unmanned Spy Plane" *Los Angeles Times*, January 8, 2001.

Proctor, Paul, "Sensor Deprivation," *Aviation Week & Space Technology*, January 31, 2000, p. 17.

Pugliese, David, "Canada Delays UAV Acquisition," *Defense News*, June 14, 1999, p. 18.

Ricks, Thomas E., and Anne Marie Squeo, "The Price of Power: Why the Pentagon Is Often Slow to Pursue Promising New Weapons," *Wall Street Journal*, October 12, 1999, p. 1.

Scott, William B., "F-22 Flight Tests Paced by Aircraft Availability," *Aviation Week & Technology*, October 16, 2000, pp. 53–54.

Smith, Giles K., "The Use of Flight Test Results in Support of High-Rate Production Go-Ahead for the B-2 Bomber," internal docment, Santa Monica: RAND, 1991.

Smith, Giles K., "Use of Flight Test Results in Support of F-22 Production Decision," internal document, Santa Monica: RAND, 1994.

Smith, Giles K., A. A. Barbour, Thomas L. McNaugher, Michael D. Rich, and William L. Stanley, *The Use of Prototypes in Weapon System Development*, R-2345-AF, Santa Monica: RAND, 1981.

Smith, Giles K., Hyman L. Shulman, and Robert S. Leonard, *Application of F-117 Acquisition Strategy to Other Programs in the New Aquisition Environment*, MR-749-AF, Santa Monica: RAND, 1996.

Sommer, Geoffrey, Giles K. Smith, John L. Birkler, and James R. Chiesa, *The Global Hawk Unmanned Aerial Vehicle Acquisition Process: A Summary of Phase I Experience*, MR-809-DARPA, Santa Monica: RAND, 1997.

Thirtle, Michael R., "Origination of the High Altitude Endurance (HAE) UAV ACTD," internal document, Santa Monica: RAND, 1998.

Thirtle, Michael R., Robert V. Johnson, and John L. Birkler, *The Predator ACTD: A Case Study for Transition Planning to the Formal Acquisition Process*, MR-899-OSD, Santa Monica: RAND, 1997.

U.S. General Accounting Office, "Best Practices: Better Management of Technology Development Can Improve Weapon System Outcomes," GAO/NSIAD-99-162, July 1999.

U.S. General Accounting Office, "Defense Acquisition: Advanced Concept Technology Demonstration Program Can Be Improved," GAO/NSIAD-99-4, October 1998.

U.S. General Accounting Office, "Unmanned Aerial Vehicles: DOD's Demonstration Approach Has Improved Project Outcomes," GAO/NSIAD-99-3, August 1999.

U.S. General Accounting Office, "Unmanned Aerial Vehicles: Progress of the Global Hawk Advanced Concept Technology Demonstration," GAO/NSIAD-00-78, April 2000.

U.S. General Accounting Office, "Unmanned Aerial Vehicles: Progress Toward Meeting High Altitude Endurance Aircraft Price Goals," GAO/NSIAD-99-29, December 1998.

Wall, Robert, "U.S. Surveillance Aircraft to Get Budget Boost," *Aviation Week & Space Technology*, January 31, 2000, pp. 32–33.

Wall, Robert, "USAF Maps Out Future of Global Hawk UAV," *Aviation Week & Space Technology*, July 12, 1999, p. 53.

Zaloga, Steven J., "Conflicts Underscore UAV Value, Vulnerability," *Aviation Week & Space Technology*, January 17, 2000, pp. 103–112.

OTHER PROGRAM DOCUMENTATION

Air Force Operational Test and Evaluation Center, Detachment 1, *Global Hawk System Advanced Concept Technology Demonstration: Quick Look Report for Roving Sands '99*, 1999.

Basic Systems for the High Altitude Endurance Unmanned Aerial Vehicle System (HAE UAV): ACAT Level I, Initial Requirements Document CAF 353-92-II, 1999.

Defense Advanced Research Projects Agency, *Advanced Concept and Technology Demonstration (ACTD) Management Plan (MP) for the Medium Altitude Endurance (MAE) Unmanned Aerial Vehicle (UAV)*, Arlington, VA, 1994.

Defense Advanced Research Projects Agency, *HAE UAV ACTD Management Plan*, Arlington, VA, December 15, 1994.

Defense Advanced Research Projects Agency, *High Altitude Endurance Unmanned Aerial Vehicle: Advanced Concept Technology Demonstration: Management Plan*, Arlington, VA, 1994.

Defense Advanced Research Projects Agency, *High Altitude Endurance Unmanned Aerial Vehicle Program (HAE UAV): Advanced Concept Technology Demonstration: Management Plan*, Arlington, VA, 1997.

Defense Advanced Research Projects Agency, *System Specification for the Global Hawk High Altitude Endurance (HAE) Unmanned Air Vehicle*, Arlington, VA, 367-0000-003E, 1997.

Demonstration & Evaluation Integrated Process Team Operations Plan, 1997.

Director of Systems Acquisition for Under Secretary of Defense (Acquisition & Technology) and Principal Deputy Under Secretary of Defense (Acquisition & Technology), Executive Summary regarding the Global Hawk DAE Review Memorandum, 1999.

Global Hawk Program Office, ASC/RAV, *Global Hawk Program Monthly Acquisition Reports*, September–December 2000.

Global Hawk Program Office, ASC/RAV, *Global Hawk Program Quads*, August–November 2000.

Global Hawk Program Office, ASC/RAV, *Global Hawk Program Single Acquisition Management Plan*, November 1, 2000.

Global Hawk Program Office, ASC/RAV, *Global Hawk System Program Overview*, December 14, 2000.

Global Hawk Program Office, ASC/RAV, *Global Hawk Update to Dr. Gansler, SAF/AQIJ*, 1999.

Heber, Charles E., Jr., *Air Combat Command's Role in the HAE UAV ACTD*, HAE UAV Program Office, 1998.

High Altitude (HAE) Unmanned Aerial Vehicle (UAV) Advanced Concept Technology Demonstration (ACTD), PMD 2404 (1)/PE# 35205F, 1999.

High Altitude Endurance Unmanned Aerial Vehicle Program Office—Aeronautical Systems Center, *Joint High Altitude Endurance Unmanned Aerial Vehicle Program (HAE UAV) Single Acquisition Management Plan*, Wright-Patterson Air Force Base, OH, 1999.

Hooper, Major Guy, handouts from presentation entitled "Global Hawk ACTD Lessons Learned," 2000.

McPherson, Colonel Craig, Global Hawk Program Director, ASC/RAV, presentation entitled "Global Hawk System: Early Strategy and Issues Session," 2000.

McPherson, Colonel Craig, Global Hawk Program Director, presentation for congressional staffers entitled "Global Hawk System," 2000.

McPherson, Craig, *RQ-4A Global Hawk: High Altitude Endurance Unmanned Aerial Reconnaissance System: Command, Control, Communications, Computers and Intelligence Support Plan*, Wright-Patterson Air Force Base, OH: Global Hawk Program Office—Aeronautical Systems Center, 2000.

Memorandum for ASC/RA from HQ AFMC/DO, Wright-Patterson Air Force Base, OH, regarding Responsible Test Organization (RTO) designation, Global Hawk System Test Program, 2000.

Memorandum for ASC/CC from SAF/AQ regarding Dark Star termination, 1999.

Memorandum for Record from Office of the Under Secretary of Defense regarding High Altitude Endurance (HAE) UAV Program Review, 1999.

Memorandum for the Secretary of the Air Force (ATTN: Acquisition Executive) from the Under Secretary of Defense regarding Global Hawk Decision Memorandum, 1999.

Memorandum for Under Secretary of Defense (Acquisition & Technology), Principal Deputy Under Secretary of Defense (Acquisition & Technology), regarding the Read-Ahead Global Hawk DAE meeting, July 7, 1999.

Sullivan, Kevin J., *High Altitude Endurance Unmanned Aerial Vehicle (HAE UAV) Monthly Acquisition Report*, 1999.

Teledyne Ryan Aeronautical, *Air Worthiness Five (5) Through Air Worthiness Seven (7) Test Missions: Detailed Test Plan for the Tier II Plus High Altitude Endurance Unmanned Aerial Vehicle System*, San Diego, CA, Report No. TRA-367-5000-157, 1998.

Teledyne Ryan Aeronautical, *Master Test Plan for the Tier II Plus High Altitude Endurance (HAE) Unmanned Air Vehicle*, San Diego, CA, Report No. TRA-367-5000-67-R-001, 1995.

Teledyne Ryan Aeronautical, *Master Test Plan for the Tier II Plus High Altitude Endurance (HAE) Unmanned Aerial Vehicle*, San Diego, CA, Report No. TRA-367-5000-67-R-001A, November 17, 1995.

Teledyne Ryan Aeronautical, *Payload Test Missions*, San Diego, CA, Report No. TRA-367-5000-202, 1998.

U.S. Atlantic Command, *HAE UAV ACTD Integrated Assessment Plan*, 1998.

U.S. Atlantic Command, *HAE UAV ACTD Joint Concept of Operations*, 1996.

U.S. Atlantic Command, *HAE UAV Joint Employment Concept of Operations*, 1998.

U.S. Joint Forces Command, *Global Hawk System Advanced Concept Technology Demonstration: Military Utility Assessment*, April 1995 to June 2000.